MANAGING TECHNOLOGICAL INNOVATION

Tools and Methods

World Scientific Series in R&D Management

ISSN:

Series Editor: Tugrul U Daim *(Portland State University, USA)*

Published

Vol. 1 Managing Technological Innovation: Tools and Methods
edited by Tugrul U Daim

World Scientific Series in R&D Management – Vol. 1

MANAGING TECHNOLOGICAL INNOVATION

Tools and Methods

Tugrul U Daim

Portland State University, USA

World Scientific

NEW JERSEY • LONDON • SINGAPORE • BEIJING • SHANGHAI • HONG KONG • TAIPEI • CHENNAI • TOKYO

Published by

World Scientific Publishing Co. Pte. Ltd.
5 Toh Tuck Link, Singapore 596224
USA office: 27 Warren Street, Suite 401-402, Hackensack, NJ 07601
UK office: 57 Shelton Street, Covent Garden, London WC2H 9HE

Library of Congress Cataloging-in-Publication Data
Names: Daim, Tugrul Unsal, 1967– editor.
Title: Managing technological innovation : tools and methods / Tugrul U Daim, editor, Portland State University, USA.
Description: New Jersey : World Scientific, [2017] | Series: World Scientific series in R&D management | Includes bibliographical references.
Identifiers: LCCN 2016059241 | ISBN 9789813207264 (hardback : alk. paper)
Subjects: LCSH: Technological innovations--Management.
Classification: LCC HD45 .M3275 2017 | DDC 658.5/77--dc23
LC record available at https://lccn.loc.gov/2016059241

British Library Cataloguing-in-Publication Data
A catalogue record for this book is available from the British Library.

Desk Editor: Amanda Yun

Typeset by Stallion Press
Email: enquiries@stallionpress.com

Printed in Singapore

Preface

Managing Technological Innovation: Cases and Tools for Managing Research and Development

The only concept that is not impacted by change around technologies is the fact that there will always be changes. This requires, as a result, all professionals dealing with technologies to be able to manage ambiguity and uncertainty, because managing the technological innovation process requires the mastery of these skills due to the complexity of the surrounding systems. After all, research and development (R&D) is one of the most important functions of organizations, and is where technological innovations are developed and launched from. This book is intended to provide those managing R&D organizations with the tools necessary to navigate technological innovations more effectively.

To develop a good set of tools, I invited authors who have published papers in the *International Journal of Innovation and Technology Management* that demonstrate the use of an effective method or tool, to contribute to this volume. The chapters in these books were developed based on these prior published papers, and revised with practicing R&D managers.

This book has five parts. The first chapter, which focuses on policies, makes up part one of this volume. Policies or strategies provide direction to R&D organizations and are therefore extremely important in managing the technological innovation.

The next four chapters (Chapters 2–5) form part two and focus on technology assessment, and provide methods and cases on evaluating technologies. Technology assessment provides evaluation platforms where technologies matching strategic directions are identified. The next four chapters (Chapters 6–9) form part three, and focus on the development of portfolios which should cover multiple objectives and mitigation strategies. In this part, analyses from technology assessment are used to build portfolios of technological innovations in the form of new products. The next three chapters (Chapters 10–12) form part four and focus on new product development, which is the execution phase for the portfolios built in the prior stage. The last two chapters (Chapters 13 and 14) form part five and focus on technology diffusion. It is critical to understand the factors which impact the diffusion of technological innovations so that they are able to diffuse into the market place at a faster rate.

As a whole, this book provides a set of cases and tools for R&D managers emphasizing important areas to pay attention to.

Tugrul U Daim, PhD and PICMET Fellow
Professor and Director of Technology Management Doctoral Program
Department of Engineering and Technology Management
Portland State University

Contents

Chapter 1

Technology Policy: Determining Effects of Incentives for Industry Competitiveness Using System Dynamics

Martin Kaggwa, Jasper L. Steyn† & Anastassios Pouris‡*

In an effort to support competitiveness in domestic industries, many late developing countries offer incentives based on investment in machinery and tooling, and local R&D. The configuration of the incentives is often based on internalized judgment rather than on formal models, making it difficult to assess such interventions objectively and to improve on them. Formalization of such policy interventions makes explicit the factors underlying performance and can reveal high-leverage policy options available to policy makers to influence performance. This chapter is aimed at showing how formalization of incentives policy for competitiveness can reveal insights useful in understanding performance of the target industry and how it can guide subsequent improvement of such policy interventions. This was done using the case of South Africa's automotive industry.

*Sam Tambani Research Institute Alris Building, Johannesburg, South Africa.
†Graduate School of Technology Management, University of Pretoria, Pretoria, South Africa.
‡Institute of Technological Innovation, University of Pretoria, Pretoria, South Africa.

Like in many developing countries, South Africa's adoption of an outward-looking industrial development policy strategy after its 1994 democratic transition was motivated, in part, by the desire to detach domestic industry performance from national economic growth (Black, 2001). It was acknowledged that the domestic market was not able to support sufficiently high production volumes that could allow efficient and competitive domestic production. Given the emphasis put on exports and foreign investment to drive national growth, international competitiveness became an important component of overall national development strategy. South Africa's policy makers hoped to emulate the successful interventions of some East Asian governments that had achieved high economic growth rates through exporting (Edwards and Golub, 2004). In 1995, the South African government launched a Motor Industry Development Program (MIDP) aimed at establishing a competitive industry, both locally and globally. Under the program, government provided the industry with import duty rebates based on local content exported and with a duty free allowance. The MIDP replaced a series of protection measures and local content requirements that had previously characterized the industry (Black, 2001). The main objectives of the MIDP were to increase competitiveness of the industry, encourage industry growth through export, stabilize employment levels, and improve the industry's trade balance and to make vehicles more affordable in the domestic market (Barnes and Black, 2003). In 2000, government introduced an investment incentive for the industry, the Productive Asset Allowance (PAA). The PAA was intended to support further efforts to make the domestic industry competitive in the long term. Investment qualifying for the PAA was widely defined to include capitalized R&D expenditure (Kaggwa *et al.*, 2007).

Despite government's offer of substantial incentives to the local automotive manufacturing industry, investment in R&D remained minimal. For the first decade of the incentives offer,

1995–2004, R&D investment as a proportion of total industry investment was less than 10% on average (Table 1). These low levels are corroborated by R&D expenditure of only 0.8% of sales in the general machinery and equipment sector (Oerlemans *et al.*, 2003). Investment in R&D in South Africa's automotive sector continued to decline post 2004. The average portion of total investment in the sector devoted to R&D was a meager 6% in the second decade, 2005–2014 (NAAMSA, 2015).

The nature of investment realized under the MIDP was an issue of concern in as far as the industry competitiveness objective was concerned since R&D investment, as opposed to capital investment, is associated with improved industry competitiveness (Stumpf and Vermaak, 1996).

Investment in plant, machinery, and tooling is important in the realization of short- to medium-term profitability of firms,

Table 1: Investment expenditure by South African vehicle manufacturers.

Year	Total Investment (Rm)	Investment in Support Infrastructure (Incl. R&D) as a % of Total Investment	Investment in Plant, Machinery and Tooling as a % of Total Investment
1995	847	9.2	86.6
1996	1171	11.1	85.0
1997	1265	8.8	81.0
1998	1342	10.4	85.2
1999	1511	7.6	87.0
2000	1562	9.0	83.9
2001	2078	11.8	86.6
2002	2726	9.6	84.8
2003	2325	8.3	85.5
2004	3577	10.1	86.9

Source: Department of Trade and Industry South Africa (2004) and NAAMSA Annual Report (2001, 2006, p. 15).

but in the long run it is the R&D investment and the subsequent potential to innovate that is likely to determine industry competitiveness (Fan, 2006; Özçelik and Taymaz, 2004; Lee, 2000; Koschatzky *et al.*, 2001).

This chapter presents a system dynamics (SD) model as a proposed instrument in formalizing the offer of incentives, applied to the South African government's offer of incentives to the automotive manufacturing sector. The model was developed from qualitative and quantitative information on how the incentives had been structured. It was used to test how changes in the incentive policy rules affected industry performance dynamics pertaining to progress toward competitiveness.

The Link between R&D Investment and Industry Competitiveness

R&D can be defined as a formal improvement-driven undertaking to discover new knowledge about products, processes, and services. It comprises of the bulk of creative systematic activities undertaken to increase the stock of knowledge and the subsequent use of this knowledge to devise new applications (Fang-Ming *et al.*, 2009). According to Zhouying (2005), R&D entails developing of technologies that can be commercialized under independent intellectual property rights. R&D is seen as the foundation of technology progress and sustainable competitiveness in the modern era (Frankema and Lindblad, 2006; Wint, 1988; Lim, 1994; Lengnick-Hall, 1992; Solow, 1957).

Competitiveness, on the other hand, refers to the ability of a firm or industry to increase in size, market share, and profitability. Quoting the US Presidential Commission on Industrial Competitiveness, Clark and Guy (2000) define competitiveness as "the degree to which it (a nation) can, under free and fair market conditions, produce goods and services that meet the test of international markets while simultaneously maintaining and expanding the real income of citizens".

The contemporary thinking is that the link between R&D and competitiveness is via its effect on technological development and subsequent innovation. Innovation, technological advances, and country competitive advantage happen to be connected by complex multidimensional relationships (Lengnick-Hall, 1992). It is reasoned that competitiveness depends on average production costs. Production costs are a function of price and nonprice factors, some of which are R&D capabilities and the ability to adopt and use new technologies. Sustainable competitiveness depends on the ability of a country or industry to offer comparable products to its competitors at lower prices on an open market. This requires that a country or industry is able to lower its production costs without sacrificing quality. Technology innovation offers one of the most practical ways to reduce production costs while at the same time maintaining or even increasing product quality. R&D investment has a powerful positive correlation with industrial profitability, product quality, and return on investment, hence overall competitiveness (Merrifield, 1989). R&D activities generate knowledge, which is a factor of production, as such, an indirect input in the neo-classical production function (Özçelik and Taymaz, 2004). Therefore, there is general agreement that countries seeking to enhance their international competitiveness, have to engage in domestic R&D and subsequent innovative activities (Kuen-Hung, 2004; Wint, 1988).

Nonetheless, the link between R&D effort, innovation, technical progress, and competitiveness has to be qualified; it is not straightforward and is characterized by time lags. For competitiveness to be realized, R&D generated knowledge has to be adopted and commercialized by industry; otherwise, the knowledge remains valueless. R&D is an input in the long process of achieving competitiveness. Like any other input in a chain of interrelated activities of a system, the relationship between input and output may be hard to establish. One has to consider time lags and control for other "competitiveness-determining" factors that simultaneously change with R&D efforts over time.

Another challenge in the R&D and competitiveness analysis relates to measuring the effectiveness of R&D. Frankema and Lindblad (2006) point out that "figures on R&D activities and numbers of people employed in R&D activities, the commonly used indicators of R&D activity, merely inform us about the scope of efforts and financial commitments but do not offer insight into the effectiveness of R&D efforts". The R&D success rate is dependent on a range of intermediary factors such as knowledge management, technology absorptive capacity of the environment, and other soft technological variables. Zhouying (2005) claims that soft technological factors that relate to the emergence of new business technologies and cultures, such as modern management techniques, venture capital, virtual technology, incubators, and so on, constitute soft technology that provides an environment for innovation and effective application of technologies, hence attainment of competitiveness. Because of the complex relationship between R&D investment and competitiveness, governments ought to carefully develop evaluation criteria for R&D sponsored programs in order to direct the behavior of recipient firms (Fang-Ming *et al.*, 2009).

Notwithstanding concerns on R&D effort and competitiveness, there is no doubt that new knowledge drives innovation and new knowledge is rooted in R&D activities. Innovation and technological capability are important assets for any country or industry in getting a competitive edge over its rivals in free and contestable markets. Achieving competitiveness is closely related to and intertwined with technology progress. Stumpf and Vermaak (1996) emphasize that global competitiveness is inseparably linked to productivity improvement and technology upgrade. Carayannis and Roy (2000) on the other hand postulate that a firm's long-term competitiveness is directly proportional to its speed and acceleration of innovation. Global technology improvement has led to a decrease in product life cycles. Facilities, equipment, and worker skills are rendered obsolete long before their useful lives have been realized

(Merrifield, 1989). In order to remain competitive, firms have to innovate continuously and need to ensure that they realize a positive return to innovation-related investment over shorter periods.

Firm expenditure on R&D and innovation activities is a long-term, high-risk form of investment but one that is necessary for industrial survival and profitable growth (Merrifield, 1989; Papadakis, 1995). Because R&D and subsequent innovative activities often requires substantial investment with high risk, governments have to play a key role in encouraging and facilitating this type of investment (Frankema and Lindblad, 2006). One of the ways for governments to do this is through the offer of industry incentives.

The beginning of the 21st century has seen multinational corporations (MNCs) increasingly globalizing their R&D activities by taking advantage of governments' support mechanisms (Gupta and Godindarajan, 2000; Iwata *et al.*, 2006). The phenomenon has motivated more developing countries to introduce incentives and institutional arrangements that can attract investment, including R&D activities, in local economies. This is done with the expectation that government R&D incentives will have a positive effect on a country's innovation effort and subsequently on its competitiveness (Herrera and Nieto, 2008).

Although empirical evidence on the effect of government incentives and private sector investment in R&D is still inconclusive (Jenkins and Thomas, 2002), for the South African automotive industry, with a location disadvantage, it was unlikely that the country could attract significant investment in R&D without some sort of compensatory incentives. Against this background, the South African government introduced automotive incentives for the local industry. The PAA in particular targeted state-of-art asset investment, R&D, and capitalized expenditure on technical expertise. The expectation was that the incentives would encourage, fairly equally, all the forms of investment, and ultimately contribute toward efforts to make the domestic industry competitive in the long term. More than 10 years after the commencement of the incentives, most of the investment realized in South Africa's automotive sector has been in form of plant, machinery, and equipment rather than R&D.

The ineffectiveness of South Africa's automotive manufacturing incentives to direct investment in R&D and subsequent innovative activities necessitated a review of the automotive industry incentives model. For this to be done rigorously, however, a formal model had to exist. Like many policies in developing countries, the MIDP policy framework was based on judgment or intuition and on consensus among stakeholders. Its assumptions remained embedded in the mental models of its historical promoters, making it hard to discern internal inconsistencies. The problem with intuitive models is that they cannot be assessed scientifically to allow objective analysis and improvement (Sterman, 2000; Forrester, 1961). In order to investigate how a change in incentive policy rules would affect industry performance dynamics in respect of the industry's competitiveness objective, an SD model of the industry incentives was developed. Formalizing of intuitive mental models has a potential to expose flaws in policy conception and can reveal the effectiveness of policy levers in the model structure (Sterman, 2000). As a result, model formalization enhances models' quality and increases the reliability of their simulations, an aspect critically important for improvement of policy intervention (Richardson and Pugh, 1981). The developed model provided a formal means to test how policy decisions on the MIDP incentives could affect industry competitiveness in the long term.

Methodology and Data

In the formalization of the competitiveness incentives for South Africa's automotive manufacturing industry, an SD approach was used. The choice of SD was motivated, in part, by the complexity of industry performance dynamics and by the need to account for feedback effects of any industry intervention.

The workings of the MIDP incentives exhibit interrelationships between industry sectors and industry performance variables without explicit cause and effect characteristic of a complex system. External inputs to the industry, like incentives, cannot be directly mapped to outcomes realized because of the complex interrelationships within the industry. Vennix (1996) contends that the SD approach is more suited to capture such relationships and feedback effects within a set of interrelated activities and processes. According to Barlas (2002) and Randers (1980), the SD approach is more applicable to policy problems where the dynamics have to do with the internal structure of the system as it has a potential to generate understanding of how a complex system is structured and can influence the effectiveness of policy input.

It was further acknowledged that policy work is often too complicated to be reduced to definite natural science laws and econometric models (Sterman, 2002). Economic policies pursued by different countries are unique and are a function of a set of circumstances peculiar to a country. They are intended to meet sets of objectives that cannot be reduced into functions to be optimized. Moreover, world circumstances are always changing, therefore, replication and generalization of particular research findings, the two strongest arguments for most dynamic optimization and pure econometric approaches, become less achievable. The key question(s) to be answered by the study were "what if" questions, that is, what would happen to industry performance in relation to the competitiveness objective if the incentive policy rules were changed. The SD simulation models are more suited to answer such "what if" questions (Richmond, 2004).

The SD as a methodology is grounded in control theory and theory of nonlinear dynamics (Sterman, 2000). The approach provides a means to capture complex relationships and feedback effects within a set of interrelated activities and processes that often characterize policy models (Vennix, 1996). It allows simulation of model performance under different policy rules giving some indication into potential outcomes of particular policy decisions before their implementation. Its presentation has a user-friendly interface that encourages users in practice to internalize the logic behind the model. In addition, the approach allows the use of quantitative and qualitative data; hence, it is not limited in its use when quantitative data are unavailable. Specialized software in SD modeling allows scenario simulations, in fairly easy and understandable steps, an aspect that is important in practice and in applied research. For this project Stella software version 9.0.2 was used, as developed by the company isee systems (formerly High Performance Systems — www.iseesystems.com).

Quantitative industry performance data for estimation of model parameters and rates of change were collected from the Department of Trade and Industry South Africa, and from the National Association of Automotive Manufacturers of South Africa (NAAMSA) publications.

One of the biggest challenges in the formalizing of a policy is access to relevant qualitative data that capture the thinking behind a policy. Developing the MIDP incentive model required specific understanding of the intentions of the program promoters and the assumptions underlying the dispensation. Such data were not explicit in the numerical and written data sources. It is widely acknowledged by a number of leading authors on SD modeling that the most important data required for building an SD model is often qualitative (Luna-Reyes and Anderson, 2003). Specific to policy work, subsequent formulation of a dynamic hypothesis and formulation of a qualitative model requires insight into the mental models of role players (Morecroft, 1985; Sterman, 2002). Key variables underlying behavior of interest resides in the mental database of some of the actors (Randers, 1980). Sterman (2002) asserts that more often mental data cannot be accessed directly but must be elicited through interviews, observation, and other methods. In order to access qualitative data, the researcher usually needs to interact with people involved in the study situation over and above the use of archival research, data collection, interviews, and direct observation or participation. The research had to include means to tap into information in stakeholders' recollection of the considerations that had led to particular policy decisions. In this regard, qualitative data to support the model building process were collected using two techniques: participant observation and discourse analysis. The researcher attended MIDP policy review meetings for a period 18 months. This was followed by discourse analysis of the qualitative information gathered from observation. In the following section, the development of the model is explained via a stepwise increase in scope.

PAA Model Structure Equations

The model was constructed sequentially for the two main incentives for automotive manufacturing in South Africa as of 2012, the PAA and the import–export complementation (IEC) scheme. The PAA incentive allowed receipt of import duty rebates based on investment, whereas the IEC allowed receipt of duty offsets based on the value of exported local content (ITAC, 2005).

Under the PAA, only investment in new and unused productive assets qualified for benefit (Kaggwa *et al.*, 2007). The value of assets qualifying for the PAA was therefore a proportion of total industry investment that was captured by the equation:

$$P_{AA}I_t = \alpha I_t \tag{1}$$

where $P_{AA}I_t$ is the PAA qualifying investment in year t, α is the PAA qualifying investment fraction, and I_t is total annual industry investment.

Benefit from the qualifying investment took the form of import duty rebates and was set at 20% of the qualifying investment. The value of rebates that could be generated from a particular value of qualifying investment was presented as:

$$P_{AA}RG = 0.2P_{AA}I_t \tag{2}$$

where $P_{AA}RG$ is the PAA rebate value generated per annum and the 0.2 is the existing PAA benefit fraction.

Since the benefit from the PAA was spread over a 5-year period, the value of annual rebate certificates generated per year was according to the equation:

$$\mathrm{RCR} = P_{AA}RG/5 \tag{3}$$

where RCR represents the value of rebate certificate release per year and the 5 represents the 5-year period over which the PAA benefit was spread.

The value of imports that could be brought into the country using PAA rebates depended on the prevailing import duty rate and the value of rebates issued in a particular year according to the equation:

$$P_{AA}RI = \mathrm{RCR/IMPORTDUTY} \tag{4}$$

where $P_{AA}RI$ is the value of imports that can be brought into the country, using the PAA rebates and IMPORTDUTY is the prevailing import duty rate in the year under consideration.

Next, the feedback effect of PAA rebates was incorporated. The first step was to make industry investment endogenous. This was done by introducing an investment rate variable. Industry investment per year was set to depend on annual investment rate, that is,

$$I_t = I_{t-1}(1 + I_{rate}) \tag{5}$$

where I_{rate} is the annual investment growth rate.

Equation (5) depicts, implicitly, a potential exponential increase in industry investment over time. This growth rate assumption was used as a baseline. The model allowed the introduction of more complex growth rates.

One of the most important aspects of SD modeling and a source of insights into system performance is the identification of feedback effects (Sterman, 2000). These constitute closed loops, where the level of outcomes has an effect on the level of inputs. In the PAA incentive model, two investment-determining factors were identified and introduced: domestic market and exports. Investment depends on planned production and planned production is, to a large extent, a function of projected domestic market size and export levels. Exports augment the domestic market size while imports, whether rebated or otherwise, reduce the effective domestic market. Local market size is a major factor in investment location decisions particularly in the automotive manufacturing industry. Jenkins and Thomas (2002) mention that the size of the local market is believed to the most important motivation for establishing European subsidiary companies in Southern Africa. As is to be expected, European subsidiary companies are strongly represented in South Africa's automotive manufacturing industry.

PAA rebatable imports added to the stock of industry imports into the country on which the industry did not pay duties. Given that the only way industry could benefit from the PAA incentive was through importing and offsetting duties payable using earned rebate certificates, firms had to import until they have exhausted import rebates received.

To account for the effects of domestic market size, exports, and PAA rebatable imports on investment, a normal-investment-growth fraction variable was introduced. At this stage of model construction, both domestic market and exports were taken to be static. To the extent that the above three variables affected investment, actual investment growth fraction differed from the normal growth fraction. The difference was the effect emanating from a production potential factor (the basis of domestic production plans), which was postulated to be proportional to: (domestic market + exports — PAA rebatable imports)/(domestic market + exports). The logic of the equation is that as long as there are no rebatable imports, investment will grow at a normal rate dictated by the size of the domestic market and export potential.

The effect of PAA rebatable imports on the production potential factor, which in turn affected the actual investment growth fraction, constituted a closed loop of the PAA incentive model.

By specifying initial model values and providing input values, the PAA model enabled simulation of the value of rebatable imports under different scenarios pertaining to the PAA benefit policy rule.

As previously mentioned, the import–export complementation scheme also contributed to rebatable imports. Therefore, the effect of rebatable imports on production plans was underestimated in the PAA model above, as it did not take into account additional rebatable imports generated through the IEC incentive scheme. Hence, the model was extended to include the IEC, which also contributed to the value of rebatable imports for the industry.

Import–Export Complementation Model Structure Equations

Under the import–export complementation, firms earned Import Rebate Credit Certificates (IRCCs), based on a proportion of exported local content. Exports were specified as being determined by an export growth rate, which rate was assumed to be determined exogenously; locally based vehicle assemblers' exports are largely dependent on parent company decisions. As such, the equation for industry export value per year was presented as

$$E_t = E_{t-1}(1+\beta) \tag{6}$$

where E_t is total industry export value per annum in year t, and β is the export growth rate fraction.

The exported value of local content was captured by the equation:

$$\text{ELC} = \text{ELCF} \times E_t \tag{7}$$

where ELC is the exported value of local content and ELCF is the exported local content fraction.

In calculating the IRCC value to be awarded to an exporting entity, the exported value of local content was discounted at a rate determined by government. The IRCC value generated, therefore, was a function of exported local content and the exported local content beneficiation fraction as determined for a particular year. Equation (8) captures this relationship:

$$\text{IRCCVALUE} = \text{ELC} \times \text{ELCBF} \tag{8}$$

where IRCCVALUE is the value of IRCCs generated per year, and ELCBF is the export local content beneficiation fraction.

By specification, the value of rebatable imports generated under the IEC in a particular year was equivalent to the value of IRCCs issued and was independent of the import duty rate.

Given this further adjustment, IRCCs generated under the IEC added to the overall stock of industry rebatable imports. To estimate the overall effect of rebatable imports on production plans, the PAA model and the IEC model were combined. A new variable, namely industry rebatable imports, which was a summation of PAA rebatable imports and IRCC rebatable imports was introduced. The direct link between PAA rebatable imports and production plans was removed and instead a link between PAA rebatable imports and industry rebatable imports on one hand, and IRCC rebatable imports and industry rebatable imports on the other, was created. Thereafter, industry rebatable imports were linked to the production potential factor. An important aspect to take note of, under the combined PAA-IEC model, is the fact that exports and the domestic market were allowed to vary over time through the introduction of respective projected growth rates.

For completeness of structure, the PAA-IEC model was extended to include an "industry trade balance" variable. Introducing the trade balance variable allowed sensitivity analysis of the industry trade balance account in response to policy decisions on the PAA and IEC incentives of the MIDP. Trade balance trend was assumed to be an important indicator of industry competitiveness in the international market.

The variable "Industry imports" was specified as an endogenous variable that depended on the import decision. The domestic market and the value of rebatable imports at industry level influenced the import decision. Before a firm within the industry could import, it had to have some insight into how much imports the domestic market could absorb. After establishing the import absorption capacity of the domestic market, the firm would have to consider the almost mandatory import it had to undertake in order to make use of import rebates earned. Hence, it was postulated that the domestic market and rebatable imports were the determining factors of the import decision. If there was no commensurate increase in the domestic market, there was a high likelihood that as rebatable imports increased, industry imports would also increase.

In quantifying the part of the model underlying the import decision, the impact of domestic market and rebatable imports on imports growth fraction was specified as being dependent on the ratio of industry rebatable imports and on the domestic market. This impact declined as the value of rebatable imports tended toward the domestic market size.

Model Validation and Testing

Model validation, as an important step in creating confidence in a model, was done before policy scenario simulations. It was acknowledged that despite the wide use of the word "validation" in modeling literature, models cannot be validated — if validation is taken to mean establishing truthfulness of the model. This is so because all models are simplified representations of reality developed with a mind-set biased toward what the model is intended to simulate and for whom it is intended. What can be validated, however, are the analytical statements and propositions derived from the axioms of closed logical systems (Sterman, 2000). In fact, the issue is not about the validity of the model but its usefulness. When SD practitioners refer to model validation, they are usually referring to ways to make the model useful and acceptable to the intended clientele, a process that is more subjective than scientific. Referring to model validation Forrester (1961) wrote:

> "Objective model-validation procedure rests eventually at some lower level of judgment or faith that either the procedure or its goals are acceptable without objective proof".

Model validation took the form of establishing that the developed model could, to a reasonable extent, replicate reference mode behavior of interest, which in this case it was industry vehicle manufacturers' (also referred to original equipment manufacturers [OEMs]) investment. Despite the fact that the objective of SD modeling is not point prediction of a system's performance but rather to probe dynamics underlying a particular behavior, it is important that an SD model can endogenously reproduce the reference mode of interest. Without replication of the reference mode, the model becomes irrelevant in providing insight into the problematic situation and as such cannot be useful. Richardson and Pugh (1981) state that if a model cannot reproduce its reference behavior mode, it is invalid. The simulation base run showed that the model could endogenously replicate the smoothed reference mode behavior. Replication of the reference behavior from an endogenous perspective indicated that the model could be useful in revealing leverage variables or points of action that could influence the industry investment trend.

Policy Rule Changes and Implications to Industry Competitiveness

Under the PAA, there were only two policy variables under control of government: the PAA benefit fraction and the import duty. Hence, policy decisions on the PAA related fundamentally to adjusting the PAA benefit fraction and/or industry import duties. Model simulation of vehicle manufacturers' investment revealed that a change in the PAA benefit fraction had minimal influence on the investment trend. The insensitivity of investment to PAA benefit fractions pointed to the fact that investment in the sector was more likely a function of domestic economic fundamentals, rather than of incentives as a specific policy lever.

On the other hand, a change in import duties did have a noticeable effect on OEM investment in the medium term. Model simulation showed that in 2012, investment by OEMs would be a billion rand (some USD 130 million) less if import duties were to be reduced from 30% to 5%. The inverse relationship between import duty rate and domestic investment suggested that the size of the domestic market was an important determinant of domestic investment. Lowering of import duties encouraged automotive imports which replaced local supply to the domestic market, reducing local production potential, consequently leading to less planned investment. The reduction in domestic investment because of lowering import duty also indicated, though implicitly, that the domestic industry was not competitive relative to other locations producing comparable automotive products including vehicles. Otherwise, imports would not increase, with a reduction in import duties, to the extent of causing an adjustment in domestic investment if the South African location was as competitive.

Under the IEC, policy makers had one policy lever under their control: the Exported Local Content Benefit Fraction (ELCBF). Model simulations showed that a change in ELCBF had a significant effect on investment. For example, OEM investment would reach some 9 billion rand in 2012 if ELCBF was to be reduced to 0.1, compared to less than 7 billion rand if the fraction remained at 0.9.

A significant effect on investment occurred with a concurrent increase in PAA benefit fraction and reduction in ELCBF. According to model simulations, by reducing ELCBF from 0.9 to 0.1 and increasing the PAA benefit fraction from 0.2 to 0.9, the difference in projected investment would increase by more than 3 billion rand in 2012. Again, the biggest impact on investment emanated from the ELCBF. The simulation was based on a "ceteris paribus" assumption but fundamentally it was indicative of the fact that the PAA benefit fraction and ELCBF were effective policy levers, in combination, in influencing industry investment.

An important observation from model simulations though, was that none of the policy decisions on the MIDP incentive model had a direct effect on the nature of investment. As such, the likelihood that the composition of automotive manufacturing industry investment in South Africa, as witnessed in the first 10 years of MIDP incentives, would change in the foreseeable future was minimal.

Insights and Conclusion

South Africa, being located at the southern end of the African continent, has a location disadvantage in terms of transport costs to and from major international vehicle markets. Given the oligopolistic nature of the global automotive manufacturing industry coupled with global oversupply of vehicles, countries need to provide vehicle manufacturers with extra motivation to invest in manufacturing in their respective economies. Through the offer of a package of incentives under the MIDP, investment growth in the domestic automotive manufacturing industry was attained. Model simulations showed that government could influence industry investment in general through adjusting incentive policy rules. However, the MIDP incentive model had no explicit policy levers to influence the nature of industry investment, particularly investment toward R&D as compared to investment in machinery, tooling, and equipment. To the extent that South Africa positioned the MIDP as competitiveness supporting intervention, the incentive model ought to have had a traceable effect on R&D and innovation activities in the local automotive industry. This was not the case. The MIDP incentive model did not include investment in R&D or innovative activities as one of its endogenous variables that could be influenced by policy makers. Investment in productive assets and in R&D was assumed homogenous. The implicit expectation was that the industry would invest proportionally in R&D and innovation according to levels in countries of which industry competitiveness was to be emulated. One way through which R&D investment and innovative activities could have been made endogenous in the MIDP model, would be to make expenditure associated with such activities a qualifying criterion for the incentives. Overall, the modeling of the MIDP incentive model revealed that the weakness of the program to support industry competitiveness emanated from the model conceptualization. The model lacked policy leverage points to influence the nature of investment. As a result, the model proved to be a weak policy framework for supporting long-term industry competitiveness.

For industrial policy formulators and technology managers, South Africa's automotive manufacturing incentive model demonstrates one of the limitations of selective industry policy to support competitiveness among late developing countries. Well-intentioned policy interventions aimed at supporting industry competitiveness end up not yielding envisaged outcomes as a result of failure to have inbuilt policy levers that can influence a targeted industry to behave in a way that puts it on a competitiveness path. Many developing countries appreciate the need to become competitive and put in place various forms of incentives to motivate industry efforts toward competitiveness. However, many of the supportive policy frameworks lack practical policy levers to channel investment in R&D and innovative activities. Moreover, systemic factors, interdependencies and feedback effects within the industry that have a bearing on industry behavior are often not taken into account. This makes performance outcomes less predictable. Ensuring that an incentive policy does indeed support industry competitiveness requires formal policy modeling and identifying levers within the model that can influence industry to make competitiveness-oriented decisions. System dynamic modeling provides a useful way to do this.

Specific to investment incentives, it is imperative to note that firm investment is not homogenous in as far as industry competitiveness is concerned. The nature of enabled investment by way of offer of incentives or subsidies has a bearing on a particular industry's progress toward global competitiveness. In particular, innovation-related expenditure is more correlated with attainment of competitiveness (Waddock and Graves, 1994). The effectiveness of incentives in supporting an industry's progress toward competitiveness, therefore, cannot be adjudicated without considering the nature of investment it enables. Parties involved in initiating and implementing incentives for competitiveness should consider explicitly how the proposed incentives would motivate investment that supports the competitiveness objective. This chapter demonstrates how formal modeling of an incentive policy framework, using an SD approach, can reveal whether a suggested framework has inbuilt policy levers to influence the nature of investment and hence could support industry competitiveness.

References

Barlas, Y. (2002). System dynamics: Systemic feedback modelling for policy analysis. In *Knowledge for Sustainable Development — An Insight into the Encyclopedia of Life Support Systems*. Oxford, UK: UNESCO-EOLSS Publishers.

Barnes, J. and Black, A. (2003). *Motor Industry Development Programme Review Report*. Pretoria, South Africa: Department of Trade and Industry.

Black, A. (2001). Globalisation and restructuring in the South African automotive industry. *Journal of International Development*, **13**(6), 779–796.

Carayannis, E. G. and Roy, R. I. S. (2000). David vs Goliaths in small satellite industry: The role of technological innovation dynamics in firm competitiveness. *Technovation*, **20**(6), 287–297.

Clark, J. and Guy, K. (2000). Innovation and competitiveness. *Technology Analysis and Strategic Management*, **10**(3), 363–394.

Department of Trade and Industry South Africa. (2004). *Current Developments in the South African Automotive Industry*. Pretoria, South Africa: Government Printer.

Edwards, L. and Golub, S. S. (2004). South Africa's international cost competitiveness and exports in manufacturing. *World Development*, **32**(8), 1323–1339.

Fan, P. (2006). Catching up through developing innovation capability: Evidence from China Telecom-equipment industry. *Technovation*, **26**(2), 359–368.

Fang-Ming, H., Der-Juinn, H., and Chao-Chih, H. (2009). The effect of government-sponsored R&D programmes on additionality in recipient firms in Taiwan. *Technovation*, **29**(3), 204–221.

Frankema, E. and Lindblad, J. T. (2006). Technological development and economic growth in Indonesia and Thailand since 1950. *ASEAN Economic Bulletin*, **23**(3), 303–324.

Forrester, J. W. (1961). *Industrial Dynamics*. Cambridge, MA: MIT Press.

Gupta, A. K. and Godindarajan, V. (2000). Knowledge flow within MNCs. *Strategic Management Journal*, **21**(4), 473–496.

Herrera, L. and Nieto, M. (2008). The national innovation effect according to firm location. *Technovation*, **28**(8), 540–550.

International Trade Administration Commission South Africa (ITAC). (2005). *Guidelines in Respect of Productive Asset Allowance*. Pretoria, South Africa: ITAC.

Iwata, S., Kurokawa, S., and Fujisue, K. (2006). Analysis of global R&D activities of Japanese MNCs in the US from the knowledge-based view. *IEEE Transitions on Engineering Management*, **53**(3), 361–379.

Jenkins, C. and Thomas, L. (2002). *Foreign Direct Investment in Southern Africa: Determinants, Characteristics and Implications for Economic Growth and Poverty Alleviation*. CREFSA Research Report. London: London School of Economics.

Kaggwa, M., Pouris, A., and Steyn, J. L. (2007). South Africa government's support of the automotive industry: Prospects of the productive asset allowance. *Development Southern Africa*, **24**(5), 681–691.

Koschatzky, Y. K., Bross, U., and Stanovnik, P. (2001). Development and innovation potential of Slovene manufacturing industry: Analysis of an industrial innovation survey. *Technovation*, **21**(5), 311–324.

Kuen-Hung, T. (2004). The impact of technological capability on firm performance in Taiwan's electronics industry. *Journal of High Technology Management Research*, **15**(2), 183–195.

Lee, J. (2000). Challenges of Korean technology-based ventures and governmental policies in emergent-technology sector. *Technovation*, **20**(9), 489–495.

Lengnick-Hall, C. (1992). Innovation and competitive advantage: What we know and what we need to know. *Journal of Management*, **18**(2), 399–429.

Lim, D. (1994). Explaining the growth performance of Asian developing economies. *Economic Development and Cultural Change*, **42**(4), 829–844.

Luna-Reyes, L. P. and Anderson, D. L. (2003). Collecting and analyzing qualitative data for system dynamics: Methods and models. *System Dynamics Review*, **19**(4), 272–296.

Merrifield, B. (1989). The overriding importance of R&D as it relates to industrial competitiveness. *Journal of Engineering and Technological Management*, **6**(1), 71–79.

Morecroft, J. (1985). Rationality in the analysis of behavioural simulation models. *Management Sciences*, **31**(7), 900–916.

National Association of Automobile Manufacturers of South Africa (NAAMSA) (2001). *Annual Report*. Pretoria, South Africa: NAAMSA.

National Association of Automobile Manufacturers of South Africa (2006). *Annual Report*. Pretoria, South Africa: NAAMSA.

National Association of Automobile Manufacturers of South Africa (2015). *Annual Report*. Pretoria, South Africa: NAAMSA.

Oerlemans, L. A., Pretorius, M. W., Buys, A. J., and Rooks, G. (2003). *Industrial innovation in South Africa: 1998–2000*. Pretoria, South Africa: University of Pretoria.

Özçelik, E. and Taymaz, E. (2004). Does innovation matter for international competitiveness in developing countries? The case of Turkish manufacturing industries. *Research Policy*, **33**(9), 409–424.

Papadakis, M. (1995). The delicate task of linking industrial R&D to national competitiveness. *Technovation*, **15**(9), 569–583.

Randers, J. (1980). *Elements of the System Dynamics Method*. Cambridge, MA: MIT Press.

Richardson, G. P. and Pugh, A. L (1981). *Introduction to System Dynamics Modelling with DYNAMO*. Cambridge, MA: Productivity Press.

Richmond, B. (2004). *An Introduction to Systems Thinking*. Lebanon, NH: ISEE systems.

Solow, R. M. (1957). Technical change and aggregate production function. *Review of Economics and Statistics*, **39**(3), 312–320.

Sterman, J. D. (2002). All models are wrong: Reflections on becoming a systems scientists. *System Dynamics Review*, **18**(4), 501–53.

Sterman, J. D. (2000). *Business Dynamics: Systems Thinking and Modelling for a Complex World*. New York, NY: McGraw-Hill.

Stumpf, W. E. and Vermaak, A. P. (1996). The role of technology in reconstructing South Africa's economy towards global competitiveness. *International Journal of Pressure Vessels and Piping*, **66**(1), 3–16.

Vennix, J. A. (1996). *Group Model Building: Facilitating Team Learning Using System Dynamics*. London: John Wiley & Sons.

Waddock, S. and Graves, S. (1994). Industrial performance and investment in R&D and capital goods. *Journal of High Technology Management Research*, **5**(1), 1–17.

Wint, G. W. (1998). The role of government in enhancing the competitiveness of developing economies. *International Journal of Public Sector Management*, **11**(4), 281–299.

Zhouying, J. (2005). Globalisation, technological competitiveness and the 'catch-up' challenge for developing countries: Some lessons of experience. *International Journal of Technology Management and Sustainable Development*, **4**(1), 35–46.

CHAPTER 2

Technology Assessment: Managing Risks for Disruptive Technologies

Anirban Ganguly, Roshanak Nilchiani† & John V. Farr‡*

Introduction

In the present-day business environment characterized by constant changes in consumer behavior and rapid advances in technology, the key to success of a firm lies in constant innovation and proper assessment of technologies. Inability to achieve a breakthrough innovation, coupled with an improper assessment of technology on the part of an established incumbent firm, have resulted in its eventual displacement by a new entrant bringing in a technology that rejuvenates new business opportunities (O'Connor *et al.*, 2008; Ganguly *et al.*, 2008b). The ability to innovate and properly assess a technology lies in a firm's organizational competencies (Miller and Blias, 1993). Furthermore, creativity, culture, understanding of the market, and the business model form the other key elements of an organization's ability to effectively assess a technology. Although incumbent technologies address the need of most customers and are incrementally improved along the trajectory demanded by the mainstream segment, they often

*Jindal Global Business School, O. P. Jindal Global University, Sonipat, Haryana, India.
†School of Systems & Enterprises, Stevens Institute of Technology, Hoboken, NJ, USA.
‡United States Military Academy at West Point, West Point, NY, USA.

fail to commercialize a new technology, especially if they assess it to be unappealing to the mainstream segment (Bower and Christensen, 1995; Calaway and Hamilton III, 2008; Rao *et al.*, 2006). Therefore, it is often the new market entrants who bring along with them, a breakthrough innovation that changes the whole paradigm of business opportunities (Christensen, 1997; Ehrnberg, 1995). Christensen (1997), in his widely discussed book, *The Innovators Dilemma*, illustrated this process and termed it as *disruptive technology* (DT). Over the years, the concept of DT (or disruptive innovation) has gained widespread importance both in academia as well as among practitioners. Disruptive technology is an emerging technology that manages to change the whole basis of market competition by introducing products with a different set of performance attributes — a set that is undermined by the existing customers (Bower and Christensenz, 1995; Danneels, 2004; Walsh *et al.*, 2002). Products arising out of DT target markets that are either unidentified by the existing technology or are available for completely new products (Linton, 2002). As a result, many industrial leaders are eventually displaced by the new market entrants developing a technology that is initially only viable to the niche market segment.

Since the basic functioning of a DT rests on the fact that it is initially appealing to a certain customer segment only (Christensen and Bower, 1996; Christensen, 1997; Christensen and Raynor, 2003), it produces innovations that require users/adopters to change their behavior in order to make use of the innovation (Moore, 2006. As a result, while on one hand, DTs promises a huge opportunity for a strong entry into existing and new markets, they also involve astronomical risk of failure on the other (Walsh *et al.*, 2002; Keizer and Halman, 2007). Therefore, an unsuccessful development and commercialization of a potential DT might prove to be very costly as it might fail to deliver appropriate and timely payoffs to investors (Walsh *et al.*, 2002), thereby resulting in huge losses for the organization. Since developing a "*risk free*" innovation strategy is nearly impossible in the currently prevailing business environment, identifying the possible risks and developing a mitigation strategy for each phase of the product development process are critical.

A well-developed risk assessment and mitigation strategy is essential for any engineering and technology project. One of the key aspects of any risk assessment strategy constitutes of identifying the risks associated with the project. This has been discussed in detail by researchers and practitioners both in the area of technological innovation and risk management (Barati and Mohammadi, 2008; Fan *et al.*, 2008; Fang and Liu, 2008; Lambert *et al.*, 2001; Prybylski, 2008). This is even more applicable for a DT, which consists of discontinuous and rapid development, unprecedented performance features, and their unpredictability of success (Keizer and Halman, 2007). Subsequent classification and ranking the identified risks would further enable an organization to assess their criticality and in turn develop appropriate risk mitigation strategies, thereby increasing its competitive advantage. However, there has been a void in the literature on innovation marrying DT with risk management — especially at various stages of the DT process. Additionally, there is also a dearth in the literature discussing the possible risks associated with a DT process from the viewpoint of a small start-up firm.

The aim of this chapter is to identify, cluster, and prioritize a set of risks that an organization might require to address while developing a technology that is potentially disruptive in nature. The contribution of this article to the domain of engineering and technology management is twofold. First, the chapter identifies and classifies a set of possible risks that is associated with each phase of a DT development process. Second, this chapter goes on to prioritize the risks in order to provide the relative criticality of the identified risks. Starting with an examination of the key theoretical concepts of DT and risk management, this chapter will subsequently move on to identifying a set of risks associated with the various stages/phases of a DT process. The identified risks will be subsequently clustered and prioritized in order to assess the degree of their criticality. This chapter concludes with recommendations based on the research findings along with providing possible directions for future research.

Background

Disruptive technology is a technological innovation theory that was put forward by Clayton Christensen in the nineties (Christensen, 1997). The theory was built up on a series of previous technological innovation studies (Dan and Chieh, 2008) coupled with Christensen's work on the disk drive industries. Disruptive technology can consist of a process, product, or even a business model that is more convenient to a certain customer segment (Sun *et al.*, 2008) and provides different value from the mainstream market, thereby, in the long run, destroying an incumbent organizations business model by rendering them obsolete (Dyerson and Pilkington, 2004). A DT can be a combination of a set of existing technologies or an entirely new technology by itself, whose application to problem areas might cause major paradigm shifts in technological products (Bower and Christensen, 1995; Kostoff *et al.*, 2004; Walsh and Linton, 2000). Generally, the firm introducing the DT initially concentrates on a niche portion of the market that is the least attractive to the market incumbents (Christensen, 1997; Hwang and Christensen, 2008). However, with gradual advancement in time, the technology starts improving in the value chain and gradually starts being embraced by the mainstream customers.

Products based on DT operate in a different *value network* than their existing counterparts. A value network can be defined as "the context within which a firm establishes a cost structure and operating processes and works with suppliers and channel partners in order to respond profitably to common needs of a class of customers" (Christensen and Raynor, 2003). It is always seen that the initial level of performance for a disruptive product lies well below that the value network demanded by the existing customer base, and therefore are neglected by the mainstream customer market. However, this new product/technology attracts the attention of a different set of customers — a set that were previously either under-served or over served in the market (Christensen, 1997; Christensen *et al.*, 2004; Hwang and Christensen, 2008). With gradual advancement of time, it improves its performance to a level where the mainstream market segment finds their requirement being met by the DT. The DT, with time, finally reaches a point where it surpasses the existing incumbent technology, thereby facilitating the mainstream customers to switch to the new technology. As a result, thanks to a DT, the market incumbents often find themselves bereft of customers in the long run.

One important characteristic of a DT lies in the inability to forecast its arrival (Walsh and Kirchoff, 2000). Since the disruptiveness potential of a technology can be primarily evaluated based on its characteristics mentioned in the literature (Sainio and Puumalainen, 2007), careful attention must be paid toward the various stages of a DT process — from its initiation until it captures the mainstream market. Additionally, although one of the implicit assumptions behind a DT is that will gradually diffuse into the mainstream market with time (Christensen, 1997), there has been numerous anecdotes denoting that DTs actually create altogether new industries that might replace the old markets (Walsh and Kirchhoff, 2000) — following the theory of *Creative Destruction* by Joseph Schumpeter (1942). Therefore, a proper analysis of a DT can prove to be far more lucrative for the new entrant than just gaining in market share — it can, over time, shoot the new entrant to the status of the market leader. Furthermore, historical evidence shows that a consistent pattern in business is the failure of leading companies to stay at the top of their industries when technologies or markets change, especially in the current increasingly turbulent, global, and discontinuous business world (Christensen, 1997; Thomond *et al.*, 2003). It is, therefore, imperative to provide an academically robust and a practically relevant investigation, from the viewpoint of risk management, to aid organization's ability to develop a DT as a major competitive strategy.

Risk Management

Risk can be stated as an "unwanted event that has a negative consequence" (Pfleeger, 2000). Therefore, risk management can be defined as "the process which aims to help organizations understand, evaluate, and take action on all their risks with a view to increasing the probability of their success and reducing the likelihood of failure" (Ronez, 2008). An improperly planned risk management strategy might adversely affect the successful completion of a project/process in terms of achievement of its outcomes, delivery of its outputs, time, cost, and quality (Kasap and Kaymak, 2007). As a result, organizations spend vast financial and human resources to design risk management strategies that would provide them with a competitive edge over their market rivals and minimize their financial loss.

The process of risk management consists of a series of stages that can be modified according to the need of a particular project. The overall risk management process can be divided into two broad phases — *risk analysis* and *risk management* (Chapman, 2001). Risk analysis comprises on identifying, assessing, and analyzing the risks for their criticality, while risk management is concerned with the actual progress of the risk management strategy (Chapman, 2001). These include the risk management strategy, monitoring the status of risk mitigation, and overall improvement of the strategy based on feedbacks and evaluations. Since the basic objective of this research concerns identification and assessing the risks associated with a DT development process, the portion on risk management will not be discussed as it falls beyond the boundaries of the present research.

The identification and analysis of the risks often pave the way for an effective risk mitigation strategy. Risk identification can be defined as "discovering, defining, describing, documenting and communicating risks before they become problems and adversely affect a project" (Barati and Mohammadi, 2008). A proper risk identification process ensures the identification of the widest range of risks associated with the project (Barati and Mohammadi, 2008; Kasaap and Kaymak, 2007) and its output is a list of possible risks associated with the various process stages of a project (Cooper *et al.*, 2004). Once the risks are identified, they are appropriately treated depending on the nature of the project. In the case of a simple or streamlined technology/engineering project with little uncertainty (and therefore few risks), the risks may be kept simply as a list of red flagged items (USDOT, 2006). However, for complex and highly uncertain project like developing a DT, a proper identification of the risks can serve as a very important tool for analysis, mitigation, and planning for the risks associated with the project. There is no single "best method" to identify the risks associated with a project or a technology development process. However, the two most commonly used risk identification techniques are survey interviews and brainstorming (Kasap and Kaymak, 2002). After the risks are identified, they are clustered into broad groups consisting of risks of similar nature (e.g., financial risk, schedule risk, cost risk). Classification of risks helps reduce redundancy and provides for easier management of the risks in later phases of the risk analysis process (USDOT, 2006). It also aids an organization to facilitate the process of evaluating and analyzing the risks in order to assess their criticality (as will be seen later in this research). A more elaborate discussion on the technique used in the present research is provided in the section concerning the research model and methodology.

A review of the open literatures on DT exhibited a variety of publications pertaining to the DT process and possible strategies to confront/adapt it. However, it was also noticed that there existed some gap in the available literature on DT, especially from the perspective of risk identification and management. Three important gaps that were identified by the authors' are given in the following:

- Although there has been a lot of literature discussing the DT process, no effort has been made to divide the process into a series of phases for the clarity of the readers and the practitioners.
- With the growing importance of risk management, it is important to identify the possible risks associated with a DT process. However, there is a substantial gap in the literature on DT in this matter.
- Classification and ranking a set of possible risks associated with various phases of a DT process would greatly aid an enterprise to develop a more effective risk mitigation strategy. There is a lack of open literature on this topic.

With the growing importance of risk identification and management serving an important part of an organization's operational logistics, it has become extremely important to identify and evaluate the possible risks that can be associated with any technological changes and innovations — especially those concerned with a DT process.

Stages of a DT Process

Disruptive technology can be stated as a scientific discovery that breaks through the usual product/technology capabilities and provide a basis for a new competitive paradigm (Kassicieh *et al.*, 2000). Literature on DT has repeatedly indicated that its performance trajectory is somewhat different from that of an existing technology. Therefore, based on the review of existing literature on DTs, coupled with interviews and brainstorming with subject matter experts (SMEs), we divided a typical DT process into a series of stages, which are shown in Table 1. This table contains a DT process based upon seven broad stages (phases) — starting from assessing the "Gap" in the existing market until the time when it displaces the incumbent technology.

A DT typically arrives in the market by targeting a certain segment that are either underserved or over served by the

Table 1: Stages of a DT process.

DT Process Phases	Explanation
I	Analyzing the gaps existing in customer requirements, current market, and existing technology
II	Concept development — defining the product and analyzing the research and development (R&D) capability of the firm
III	Preliminary design of a product
IV	Product and technology development
V	Preproduct launching stage — including product marketing, pricing, and preliminary assessment of the product adaptability in the target market
VI	Product launch to its target market segment
VII	Product improvement from feedbacks (as well as own R&D) in order to capture the incumbent market in the long run

existing technology. Therefore, during Phase I, a proper assessment of the existing market, along with the possible available gaps that can be targeted and fulfilled, is vital to the basic idea generation for a DT. This, therefore, forms the first stage of the process. The next few Phases (II, III, and IV) consist of developing the concept, designing the product, and developing the product/technology. This is vital to any technology development process. In the context of a DT, it is even more vital to assess the maturity level of the existing technology as a technology that has reached a certain level of maturity might have very little room for further improvement and therefore be vulnerable to competition (Ganguly *et al.*, 2007). Other important aspects of these phases comprises of initial designing of the product, developing the technology and the product evolving out of it, and so on.

Phases V and VI of the DT process comprise of the duration from the preproduct launch until the final product launch. This includes a survey of product adaptability, product marketing, and most importantly, especially in the case of a DT, pricing of the product since the savings in the cost achieved by a product arising due to a DT often proves to be one of the driving forces behind its success (Ganguly *et al.*, 2008a). A typical performance path of a DT consists of it being initially launched to a niche target market (a segment that is often ignored by the mainstream market) and consequently, with time, capturing the mainstream market segment. This is often achieved through continuous process improvement that results from feedbacks and postproduct launch activities. This, therefore, forms the final phase (VII) of a successful DT process, ultimately resulting in it displacing the incumbent technology and thereby capturing the mainstream customer segment with time.

Identification and Classification of Risks Associated with a DT Process

The process of risk analysis started with identifying a set of possible risks associated at various stages of a DT process. However, before moving on to a detailed identification of the risks, it was thought worthwhile to enlist the set of major risk categories under which the various risks will be classified. Nine major risk categories that were determined are provided in Table 2 along with their operational definition from the perspective of a DT.

Once the broad risk categories were identified, the next stage was to identify the risks associated with various stages of a DT process as shown in Table 2. The risks were identified based on the review of literature on DT and technology innovation, brainstorming, and discussion with SMEs in the domain of technology and innovation management. The risks associated with the various stages of a DT process were identified based on the following two steps:

(i) Based on the review of the literature on DT, technological innovation and technology management, the authors

Table 2: Risk categories and their operational definition from a DT perspective.

	Risk Categories	Operational Definition
1	Economic risks	The risks associated with monitory/financial loss that would result from an insufficient or failed DT process
2	Time/schedule risks	The risks associated with the likelihood of failing to achieve time/schedule deadline at various stages of a DT development life cycle and the consequence of that failure
3	Operational risks[1]	The risk or loss resulting from incomplete/inadequate/failed internal processes, including human resources and organizational dynamics, at various phases of the DT development life cycle
4	Customer risks	The risks associated with the loss in the value of the product/technology to the targeted customer segment as a result of certain changes that largely impacts the consumer behavior and market trends
5	Market risks	The loss associated due to a change in market trend and competitors activities during the development of the DT or the product arising out of it

(*Continued*)

Table 2: (*Continued*)

	Risk Categories	Operational Definition
6	Legal risks	The risks associated with uncertainty in the interpretation or application of contract, regulations, and intellectual properties that the DT might require
7	Technology risks	Likelihood of failure of a project due to activities related to technical and engineering process, engineering design, technology development, and so on.
8	Socio-political risks	The uncertainty that an organization might face due to an existing internal or external social policies and political issues as well as any sudden changes of the same
9	Quality	The risk associated with the disruptive product/technology failing to deliver the desired (or acceptable) level of quality demanded by the targeted market segment

[1] From an organizational or human resource perspectives.

held a brainstorming session in order to arrive at a set of probable risks attached to various stages of the DT process. Further research enabled the authors to reduce the initial enlisted pool to a set of important risks associated with the stages.

(ii) The identified risks were subsequently enlisted and the set was provided to a set of SMEs for their opinion and feedback. They consisted of industry professionals having an experience of 15 years or more in the field of technology and innovation management. They were chosen based on a series of criteria that were used to evaluate their knowledge in the field of technological innovation and the possible risks associated with it. Based on their opinions, a set of important risks that might be associated with a DT process is identified. This is provided in Table 2 along with their broad categories (which are exhibited in Table 3).

It should be mentioned in this context that the identified risks in Table 4 are in no capacity a holistic set of possible risks attached to a process. Rather, they are thought to be a set of relatively important risks that might occur at various phases of a process. Practitioners and enterprises assessing and developing a potential product are suggested, if required, to modify the set to suit their own business process and risk mitigation strategies.

Once the risks were identified, the next task was to classify them according to the stages that they are most likely to occur. As readers might recall, the stages of a typical process were identified earlier in the chapter and is provided in Table 1. Table 4 exhibits the seven stages of (illustrated in Table 1) and the identified risks (illustrated in Table 3) at the corresponding stages.

Table 4 illustrates the risks associated with a DT process according to the possible stages in the process that they are most likely to occur. Furthermore, certain risks (e.g., 3F and 5C) are clustered under two different phases as they can be associated with both. Finally, it should also be mentioned that although

Table 3: Possible risks associated with a DT process.

Categories		Description
Economic	***1A***	Financial risks associated with a failed concept development
	1B	Financial risk associated with an incorrect GAP analysis
	1C	Risks associated with unforeseen expenses in the design stage
	1D	Risks associated with budget overrun in order to capture the mainstream segment
	1E	Risks associated with exceeding the estimated budget for the product/technology
	1F	Financial risks associated with intellectual property (IP) infringement during the design stage of the product
Time/schedule	***2A***	Risks associated with a delay in developing the concept (someone else develops it in the meanwhile)
	2B	Wastage in time associated with a failed concept development
	2C	Risks associated with delay in developing the product — competitor might beat the start-up to the development
	2D	Delay in product launch risks
Operational[1]	***3A***	Risks associated with an improper market survey — assessment the wrong target segment
	3B	Risks associated with a wrong requirement analysis — the requirement gap assessed is different from the actual gap in requirements
	3C	Risks associated with loss of key R&D personnel during the stage of concept development
	3D	Risks associated with the concept development (e.g., selecting the wrong R&D team)

	3E	Risks associated with conflicts in corporate culture
	3F	Risks associated with outsourcing some stages of the development of the product
	3G	Loss of key suppliers during the product launch
	3H	Risks associated with outsourcing the distribution of the product
	3J	Risks associated with incorrect feedback in order to transit this technology to the mainstream market
Customer	***4A***	Risks associated with the mainstream customers not switching to the new technology/product in the long run
	4B	Failure to meet customer demands in the targeted market segment
	4C	Risks associated with rejection from the mainstream customer segment in spite of gaining a firm foothold in the niche segment
Market	***5A***	Arrival of a competitor using the same type of technology
	5B	Risks associated with unappealing marketing/advertising campaign
	5C	Risks associated with an incorrect assessment of the market growth — the actual rate of growth in the market might be slower than the estimated growth
Legal	***6A***	IP Risks in the design stage of the product
	6B	Risks associated with violation in IP in the process of developing/refining the technology/product to transit it toward the mainstream market
	6C	Risks associated with insufficient knowledge of relevant IP issues at the design and development stage

(Continued)

Table 3: (*Continued*)

Categories		Description
Technology	***7A***	Wrong initial concept — inadvertently developing an incorrect/improper technology
	7B	Risks associated with incorrect evaluation/assessment with the maturity level of the existing technology
	7C	Risks associated with wrong technology planning/product planning
	7D	Technical risks associated with product development process
Sociopolitical	***8A***	Risks associated with adverse changes in the industry regulation during the development stage
	8B	Risks associated with adverse changes to environmental regulations during the development stage
	8C	New product fails to deliver generally accepted values (health, safety, environment friendly, etc.)
Quality	***9A***	Risks associated with developing an too inferior quality product
	9B	New product fails to fulfill the intended function
	9C	The new product succeeds in the niche market but fails to deliver the expected quality demanded by the mainstream customers in the long run

[1] From an organizational or human resource perspectives.

Table 4: Disruptive technology phases and the corresponding associated risks.

DT Process Phases	Explanation	Risks
I	Analyzing the gaps existing in customer requirements, current market, and existing technology	3A, 3B, 3D, 3F, 6B
II	Concept development — defining the product and analyzing the R&D capability of the firm	1A, 2A, 3C, 3E, 3F, 6A, 6C
III	Preliminary design of a product	1B, 1C, 1F, 5A, 5C
IV	Product and technology development	1E, 2C, 3G, 5C, 6D, 7A, 7B, 8A
V	Preproduct launching stage — including product marketing, pricing, and preliminary assessment of the product adaptability in the target market	4C, 4E
VI	Product launch to its target market segment	2C, 3H, 3I, 4B, 4C, 7C, 8B
VII	Product improvement from feedbacks (as well as own R&D) in order to capture the incumbent market in the long run	1D, 3J, 4A, 4D, 5B, 8C

the risks are most likely to occur in the stages depicted in Table 4, they should be monitored at all stages of the process in case any of the aforesaid risks tends to surface in some other stages of the process.

Additionally, the basic concept of risk management indicated that risks could broadly be classified under two categories — external and internal. Therefore, it was thought pertinent to categorize the identified risks into these two categories. Figure 1 classifies the risks as either internal or external in nature.

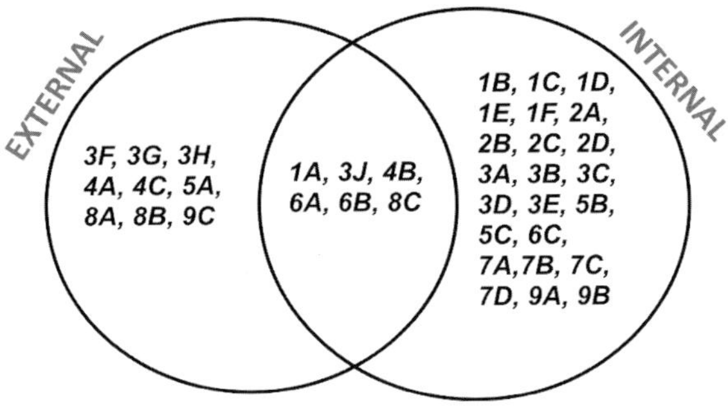

Figure 1: Risks associated with a DT process — external versus internal.

As shown in this figure, there are certain risks that can be classified either as external, internal, or a combination of both. For example, failure to meet customer demands in the target market segment can arise out of some internal failure in the organizational logistic (internal issue) or due to some unforeseen changes in the market demand (external issue). Therefore, it was thought worthy to be placed in the "intersection area" of the Venn diagram (Fig. 1). Having a knowledge regarding the external/internal nature of the risk will enable the firm pursuing the technology to have a clearer focus on the nature of the risks. This, in turn, would enable them to design their risk management plan more strategically and effectively, thereby minimizing the loss in the process.

Prioritization of the Risks to Assess Their Relative Criticality

A survey can be stated as a method of collecting information from people about their background, ideas, feelings, beliefs, and so on (Fink and Kosecoff, 1998). In the context of this research, a sample of 50 industry professionals spanning across various industrial strata were surveyed for their feedback. The surveyed professionals were well versed with the concept of technological innovation and emerging/DT. The survey conducted comprised of a structured questionnaire requesting the respondents to evaluate the likelihood of the identified risks, on a scale of ranging from "Very Low" (1) to "Very High" (5). Furthermore, they were asked to identify the potential consequence of the respective risk to a DT process. The survey feedback from the respondents was analyzed to rank and prioritize the identified risks associated with a DT process. In the context of this research, the values of the risks were calculated based on Eq. (1).

$$R_J - \frac{\sum_i^N = (p_i \times c_i)}{N} \tag{1}$$

where,

Rj = The value of the jth risk
pi = The probability of the occurrence of the jth risk as provided by the ith respondent
ci = Consequence of the jth risk as provided by the ith respondent
N = Total number of respondents

As seen in Eq. (1), the value of a particular risk is the average of the product of its probability and consequence. Since the value of the probability lies between 0 and 1, the survey responses related to the probability was translated into a scale ranging between 0 and 1. A response of "very low" was considered as having a probability 0.1 (as assigning 0 would have represented no importance of the risk) and a response of "very high" was attached a probability value of 0.9 (as assigning 1 would have represented a "complete" importance of the risk). Based on this rationale, the survey responses on the likelihood of the risks were mapped to the probability using this scale in Table 5.

Once the survey response were assigned their corresponding probability values, the next stage was to arrive at the final

Table 5: Mapping the survey response data to the risk probability.

Survey Response	1	2	3	4	5
Corresponding Probability Value	0.1	0.3	0.5	0.7	0.9

value of all of the identified risks (*Rj*, where $j = 1$–38) using Eq. (1). This is provided in Table 6 along with their standard deviation.

The value of the identified risks along with their standard deviation is provided in Table 6. The values of the standard deviation indicated that the respondents were fairly in agreement with each other, in spite of being surveyed separately, regarding the relative importance of the identified set of risks associated with a DT process. The risks were then ranked according to the descending order of their value as exhibited in Fig. 2.

As seen in Fig. 3, the risks associated with exceeding the estimated budget of the technology (1E) was identified as the most important in a DT process while the risk associated with and adverse change in the business regulation (8B) the least important. The remaining 36 identified risks were located between 1E and 8B. Furthermore, the 38 identified risks were subsequently clubbed according to their parent categories (refer Table 2 for details) and a prioritized ranking of those nine categories was performed. This is illustrated in Fig. 3.

As seen in Fig. 3, the most important category of risk facing a DT process is the one associated with time/schedule. Developing, designing, and implementing a technology at the proper time often pave the way for its success. This is even truer in the case of a DT where the market has to be monitored carefully and the timing of the development, design, and implementation perfectly executed in order to not only capture the initial niche market but also to gradually move up the curve in order to infiltrate the mainstream market segment. Furthermore, a DT initially targets a niche market with the intention of gradually

Table 6: Mean value of the identified risks.

	Description	Risk Value (R_j)	Standard Deviation
1A	Financial risks associated with a failed concept development	2.1128	0.8627
1B	Financial risk associated with an incorrect GAP analysis	2.0513	0.8476
1C	Risks associated with unforeseen expenses in the design stage	2.5410	1.0477
1D	Risks associated with budget overrun in order to capture the mainstream segment	2.3282	1.0585
1E	Risks associated with exceeding the estimated budget for the product/technology	2.6949	1.0583
1F	Financial risks associated with IP infringement during the design stage of the product	1.4872	0.9114
2A	Risks associated with a delay in developing the concept (someone else develops it in the meanwhile)	2.2692	1.0089
2B	Wastage in time associated with a failed concept development	2.1538	0.9096
2C	Risks associated with delay in developing the product — competitor might beat the start-up to the development	2.3872	0.9487
2D	Delay in product launch risks	2.3590	0.9144
3A	Risks associated with an improper market survey — assessment the wrong target segment	1.9641	1.1233

(Continued)

Table 6: *(Continued)*

	Description	Risk Value (R_j)	Standard Deviation
3B	Risks associated with a wrong requirement analysis — the requirement Gap assessed is different from the actual Gap in requirements	2.5692	1.2220
3C	Risks associated with loss of key R&D personnel during the stage of concept development	2.0359	1.1130
3D	Risks associated with the concept development (e.g., selecting the wrong R&D team)	1.9308	1.0921
3E	Risks associated with conflicts in corporate culture	1.9051	1.1720
3F	Risks associated with outsourcing some stages of the development of the product	1.5744	0.9861
3G	Loss of key suppliers during the product launch	1.4744	0.8168
3H	Risks associated with outsourcing the distribution of the product	1.2641	0.7604
3J	Risks associated with incorrect feedback in order to transit this technology to the mainstream market	2.0641	1.0366
4A	Risks associated with the mainstream customers not switching to the new technology/product in the long run	2.0436	0.9597
4B	Failure to meet customer demands in the targeted market segment	2.2026	1.0046

4C	Risks associated with rejection from the mainstream customer segment in spite of gaining a firm foothold in the niche segment	1.6385	0.7929
5A	Arrival of a competitor using the same type of technology	1.8205	1.0710
5B	Risks associated with unappealing marketing/advertising campaign	1.8462	1.0622
5C	Risk associated with an incorrect assessment of the market growth — the actual rate of growth in the market might be slower than the estimated growth	1.9615	0.9025
6A	Intellectual property risks in the design stage of the product	1.4487	0.8016
6B	Risks associated with violation in IP in the process of developing/refining the technology/product to transit it toward the mainstream market	1.4256	0.8819
6C	Risks associated with insufficient knowledge of relevant IP issues at the design and development stage	1.6462	0.8328
7A	Wrong initial concept — inadvertently developing an incorrect/improper technology	2.0385	1.0284
7B	Risks associated with incorrect evaluation/assessment with the maturity level of the existing technology	1.9641	1.0798
7C	Risks associated with wrong technology planning/product planning	1.9513	0.9912
7D	Technical risks associated with product development process	2.1179	1.0200
8A	Risks associated with adverse changes in the industry regulation during the development stage	1.3385	0.9109

(*Continued*)

Table 6: (*Continued*)

	Description	Risk Value (R_j)	Standard Deviation
8B	Risks associated with adverse changes to environmental regulations during the development stage	1.1872	0.7596
8C	New product fails to deliver generally accepted values (health, safety, environment friendly, etc.)	1.3615	0.8683
9A	Risks associated with developing an too inferior quality product	1.8974	1.1527
9B	New product fails to fulfill the intended function	1.7385	1.0153
9C	The new product succeeds in the niche market but fails to deliver the expected quality demanded by the mainstream customers in the long run	1.8513	0.9736

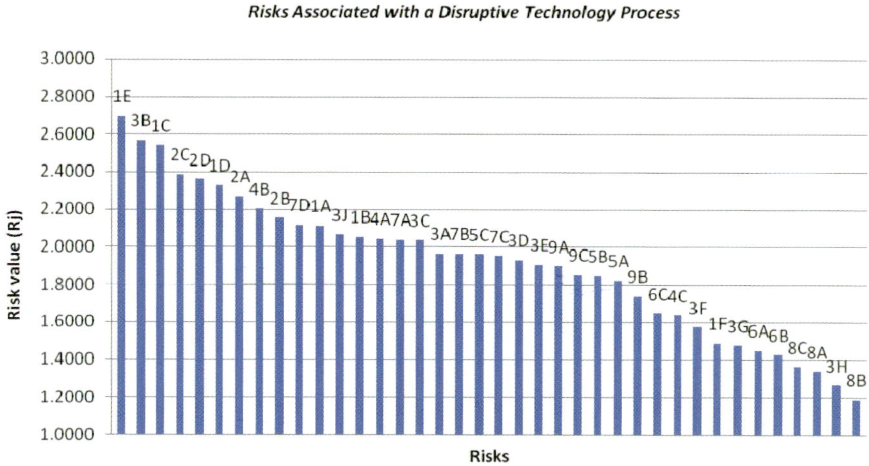

Figure 2: Ranking the individual identified risks associated with a DT process.

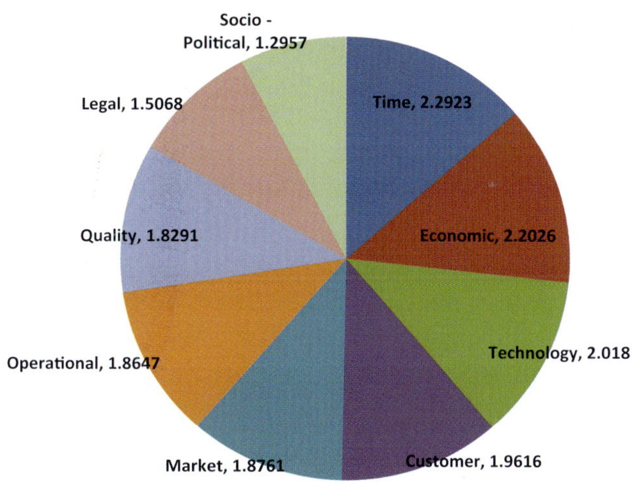

Figure 3: Ranking the overall categories of the risks associated with a DT process.

moving up to the mainstream market in the long run. In this context, it might very well happen that the technology might not be accepted in its target market segment or even if it does, fail to capture the mainstream market and eventually die down.

Either situation will endure heavy financial burden in the process — thereby illustrating the importance of the financial risks associated with the process — which comes a close second. This is followed by technology risks, customer risks, market risks, operational risks, quality risks, legal risks, and sociopolitical risks.

Conclusions

The structured process and the risk categories are important. In practice, the possible risks presented in Table 3 can serve as an initial assessment. These set can then be further modified or narrowed down to suit the specific requirements of the organization or the project. A proper identification and assessment of the risks associated with a DT process will enable an enterprise to develop a less risky strategy, thereby increasing the overall efficiency of the organization. This, in turn, will lead to a higher probability of success, all leading to achieving a greater market share and overall profit. The risks in this paper can be applied to many other technological decisions, especially as a part of the process to assess any emerging technology. It is expected that this chapter would serve as a guide for any enterprise planning to develop a risk mitigation strategy to assess an emerging technology that is potentially disruptive in nature.

This chapter also sheds some light on evaluating the relative importance of the possible risks associated with a DT process. This, along with the set of identified risks, carries important managerial implications for an organization. Managers are faced with a lot of uncertainly while evaluating an emerging technology, especially one that is potentially disruptive in nature. Therefore, having a set of possible risks associated with a DT process (as shown in Table 3) would considerably strengthen their ability to develop a more efficient and effective technology strategy. Furthermore, understanding the relative importance of the risks (as shown in Table 6) would greatly aid the managers to monitor and address the risks in a more methodical fashion. Based on their relative criticality, the managers and the decision makers would then be able to determine the risks that they need to address more emphatically.

However, the process of risk identification is a continuous one, unique for every product, and should be monitored at all times. Monitoring the risks on a regular basis would enable the managers to modify/revise the identified risks, depending upon the nature of the business environment prevailing at that point of time, to suit the requirements of the enterprise.

Future research should be directed from the viewpoint of segregating these risks into "industry specific" types in order to assess their relative priority according to the characteristics of the industries. This would provide the academicians and the practitioners with a more comprehensive analysis of the possible identified risks associated with a DT process. Additionally, analyzing the risks from the point of view of the mainstream incumbent enterprise would carry with it great potential as the findings can be used by the incumbent firms to assess an emerging technology that might be potentially disruptive in nature. Finally, developing a risk mitigation strategy based on the possible identified risks, a one that is specifically aimed toward a DT process can serve as an important area of future research and would shed considerable light in the domain of technology and innovation management.

References

Barati, S. and Mohammadi, S. (2008). Enhancing Risk Management with an Efficient Risk Identification Approach. *4th IEEE International Conference on Management of Innovation and Technology, 2008 (ICMIT 2008)*, Bangkok, Thailand, 2008, 1181–1186.

Bower, J. L. and Christensen, C. M. (1995). Disruptive Technologies: Catching the Wave. *Harv. Bus. Rev.,* **73**: 43–53.

Chapman, R. J. (2001). The Controlling Influences on Effective Risk Identification and Assessment for Construction Design Management. *Int. J. Proj. Manag.,* **19**: 147–160.

Christensen, C. M. (1997). *The Innovator's Dilemma*. New York, NY: HarperCollins Publishers.

Christensen, C. M. and Bower, J. L. (1996). Customer Power, Strategic Investment, and the Failure of Leading Firms. *Strat. Manag. J.,* **17**: 197–218.

Christensen, C. M. and Raynor, M. E. (2003). *The Innovator's Solution.* Boston, MA: Harvard Business School Press.

Christensen, C. M., Roth, E. A., and Anthony, S. D. (2004). *Seeing What's Next: Using the Theories of Innovation to Predict Industry Change.* Boston, MA: Harvard Business School Press.

Cooper, D. F., Grey, S., Walker, P., and Raymond, G. (2004). *Project Risk Management Guidelines: Managing Risk in Large Projects and Complex Procurements.* Hoboken, NJ: John Wiley & Sons.

Dan, Y. and Chieh, H. C. (2008). A Reflective Review of Disruptive Innovation Theory. *PICMET '08–2008 Portland International Conference on Management of Engineering & Technology,* Cape Town, South Africa, 2008, 402–414.

Danneels, E. (2004). Disruptive Technology Reconsidered: A Critique and Research Agenda. *J. Prod. Innov. Manag.,* **21**: 246–259.

Dyerson, R. and Pilkington, A. (2004). Expecting the Unexpected: Disruptive Technological Change Process and the Electric Vehicle. *Int. J. Innovat. Tech. Manag.,* **1**: 165–183.

Ehrnberg, E. (1995). On the Definition and Measurement of Technological Discontinuities. *Technovation,* **15**: 437–452.

Fan, M., Lin, N. P., and Sheu, C. (2008). Choosing a Project Risk-handling Strategy: An Analytical Model. *Int. J. Prod. Econ.,* **112**: 700–713.

Fang, C. L. and Liu, W. (2008). Analysis of Relationship between Operational Risk and Financial Efficiency among Different Location Types in Farmers' Association. *4th IEEE International Conference on Management of Innovation and Technology, 2008 (ICMIT 2008),* Bangkok, Thailand, 2008, 1348–1353.

Fink, A. and Kosecoff, J. (1998). *How to Conduct Surveys: A Step-by-Step Guide.* Thousand Oaks, CA: Sage Publications.

Ganguly, A., Nilchiani, R., and Farr, J. V. (2007). Using Metrics to Assess the Disruptiveness of a Technology. *28th Annual American Society of Engineering Management Conference,* Chattanooga, TN, November 7–10, 2007.

Ganguly, A., Nilchiani, R. and Farr, J. V. (2008a). Using a Systems Process to Access the Potential for a Disruptive Technology. *Conference of Systems Engineering Research (CSER) 2008,* Los Angeles, CA, April 4–5, 2008, pp. 1–10.

Ganguly, A., Nilchiani, R., and Farr J. V. (2008b). Defining a Set of Metrics to Evaluate the Potential Disruptiveness of a Technology. *Eng. Manag. J.,* **22**(1): 34–44.

Hwang, J. and Christensen, C. M. (2008). Disruptive Innovation in Health Care Delivery: A Framework for Business-Model Innovation. *Health Aff.,* **27**: 1329–1335.

Kasap, D. and Kaymak, M. (2007). Risk Identification Step of the Project Risk Management. *Portland International Center for Management of Engineering and Technology Conference (PICMET 2007)*, Portland, OR, 2007, 2116–2120.

Kassicieh, S. K., Anderson, S. K., Romig, A., Cummings, J., McWhorter, P., and Williams, D. (2000). A Model for Technology Assessment and Commercialization for Innovative Disruptive Technologies. *Proceedings of the 2000 IEEE Engineering Management Society*, Albuquerque, New Mexico, August 13–15, 340–344.

Keizer, J. A. and Halman, J. I. M. (2007). Diagonising Risks in Radical Innovation Projects. *Res. Tech. Manag.,* **50**: 30–36.

Kostoff, R. N., Boylan, R., and Simons, G. R. (2004). Disruptive Technology Roadmaps. *Tech. Forecast. Soc. Change.,* **71**: 141–159.

Lambert, J. H., Haimes, Y. Y., Li, D., Schooff, R. M., and Tulsiani, V. (2001). Identification, Ranking, and Management of Risks in a Major System Acquisition. *Reliab. Eng. Syst. Saf.,* **72**: 315–325.

Linton, J. D. (2002). Forecasting the Market Diffusion of Disruptive and Discontinuous Innovation. *IEEE Trans. Eng. Manag.,* **49**: 365–374.

Miller, R. and Blais, R. A. (1993). Modes of Innovation in Six Industrial Sectors. *IEEE Trans. Eng. Manag.,* **40**: 264–273.

Moore, G. A. (2006). *Crossing the Chasm*. New York, NY: HarperBusiness Essentials (HarperCollins).

O'Connor, G. C., Ravichandran, T., and Robeson, D. (2008). Risk Management Through Learning: Management Practices for Radical Innovation Success. *J. High Tech. Manag. Res.,* **19**: 70–82.

Pfleeger, S. L. (2000). Risky Business: What We Have Yet to Learn About Risk Management. *J. Syst. Software.,* **53**: 265–273.

Prybylski, H. (2008). Enhancing Risk Identification. *Risk Manag.,* **55**: 56–57.

Rao, B., Angelov, B., and Nov O. (2006). Fusion of Disruptive Technologies: Lessons from the Skype Case. *Eur. Manag. J.,* **24**: 174–188.

Ronez, M. Introduction to Risk Management. *Integrated Risk Management Body of Knowledge*: Asian Risk Management Institute.

http://www.arimi.org/arimi/faculty/research_BOK_RMIntro.htm# MF, Accessed on November 10, 2008.

Sainio, L. M. and Puumalainen, K. (2007). Evaluating Technology Disruptiveness in a Strategic Corporate Context: A case study. *Tech. Forecast. Soc. Change.,* **74**: 1315–1333.

Schumpeter, J. A. (1942). *Capitalism, Socialism and Democracy*. New York, NY: HarperCollins Publishers.

Sun, J., Gao, J., Yang, B., and Tan, R. (2008). Achieving Disruptive Innovation — Forecasting Potential Technologies Based upon Technical System Evolution by TRIZ. *4th IEEE International Conference on Management of Innovation and Technology, 2008 (ICMIT 2008)*, Bangkok, Thailand, 2008, 18–22.

Thomond, P., Herzberg, T., and Lettice, F. (2003). Disruptive Innovation: Removing the Innovators Dilemma — The identification of Four Critical Barriers. *British Academy of Management Annual Conference*, Harrogate, UK.

USDOT. (2006). Risk Identification. *USDOT Risk Assessment and Allocation for Highway Construction Management*, http://international.fhwa.dot.gov/riskassess/risk_hcm06_02.cfm , Accessed on November 10, 2008.

Walsh, S. and Kirchhoff, B. (2000). Disruptive Technologies: Innovators' Problem and Entrepreneurs' Opportunity. *Proceedings of the 2000 IEEE Engineering Management Society, EMS 2000*, Albuquerque, NM, 2000, 319–324.

Walsh, S. T. and Linton, J. D. (2000). Infrastructure for Emergent Industries Based on Discontinuous Innovations. *Eng. Manag. J.,* **12**: 23–31.

Walsh, S. T., Kirchhoff, B. A., and Newbert, S. (2002). Differentiating Market Strategies for Disruptive Technologies. *IEEE Trans. Eng. Manag.,* **49**: 341–351.

Chapter 3

Technology Assessment: Triz Technology System Evolution Theory

Jianguang Sun & Runhua Tan†*

Disruptive innovation (DI) is a technical innovation theory put forward by Christensen in 1997 (Christensen, 1997), and consummated by him at last (Christensen and Bower, 1996; Christensen and Overdorf, 2000; Christensen and Raynor, 2003). DI refers to a technology, process, or business model that brings to a market, a much more affordable product or service that is much simpler to use. When these products are getting stable gradually in the low-end market and in the new market by development, they can take the place of the products in the mainstream market. Though the performance of DI products is not so good when they are introduced, they have the characteristics of cheap, simple, portable, and easy to use. This makes it possible to compete with the mature products in the mainstream market.

When Christensen studied a series of initiate DI, other scholars also studied this theory all along. Dannlees thought that the concept of DI given by Christensen was not clear (Danneels, 2004). People confused the difference between DI and common new technology whose performance is under that of main technologies. Govindarajan and Kapolle believed that the cases

*Hebei University of Technology Institute of Design for Innovation, Tianjin, P. R. China.

†Hebei University of Technology Institute of Design for Innovation, Tianjin, P. R. China.

given by Christensen were all the results of the analysis after something occurred, they couldn't give beforehand forecasting (Govindarajan and Kapolle, 2006). Smith pointed out that the initial use of technology of modern computer game relative to imitating industry is a sort of low-end capacity for specific users (Smith, 2007). However, with the development of technologies, disruptive potential technique emerged. This kind of technique has already reached the level that is much more powerful in competition ability than many main facilities. It has become the technical base of the establishment of new companies. These companies may replace the ones that adopt traditional mature technologies by means of DI. The technology of DI put forward by Chieh and others should be a noteworthy direction and it should be developed as an innovation tactic (Chieh and Dan, 2007). Kohlbacher and Chieh analyzed the market of the aged using the frame of DI. They pointed out that the market supplied important commercial opportunities (Kohlbacher and Chieh, 2007). To sum up, the technology of DI has become the important investigative direction in this field.

The research based on industrial technology DI by Christensen (Christensen, 2000), by means of graphic method, can forecast the evolution of technology for the corporations in need. However, the forecasting process is difficult, in what way to forecast in general is not clear. A simple method is historical evolution trend analysis, but for new technology or new market, almost there are nonexistent historical data. The future direction of evolution is not clear that this method would not seem feasible. The present investigations suggest that there are two methods to support the technology forecasting. The first method is technology foresight. Doering and Parayre put forward a technology assessment procedure to collect, test, assess, and present for many times (Doering and Parayre, 2000). This procedure is helpful to the assessment of the future commercial value of scientific innovation. Rowe and Wright put forth Delphi technology for the acquirement of integrated technologies and market track (Rowe and Wright, 1999). The second method is technology roadmap. In recent years, technology roadmap has been made rapid progress in theory and practice, but all that are from the sustaining technology. Walsh corrected the traditional technology roadmap by using disruptive technology, constructed a commercial model of disruptive technology roadmap, and forecasted the development trend of nanotechnology (Walsh, 2004).

TRIZ is the Russian acronym for the Theory of Inventive Problem Solving (Altshuller, 1999). It mostly includes the forecast of technology maturity, technology evolution, the contradiction solution, effect, standard solution and the Algorithm for the Solution of Inventive Problems (ARIZ from the Russian abbreviation), and so forth. The computer-aided innovation (CAI) software (Kohn, *et al.*, 2005) based on TRIZ has been developed recently. All kinds of methods in TRIZ can be used either separately or together, so that different problems in invention can be solved (Tan, 2004).

This chapter is based on the technical evolution theory in TRIZ, discussed the forecasting process and method of DI, and established the process model. It offers a support in both theory and methods for forecasting a DI technology.

Theoretical Background

Christensen divides innovation into sustaining innovation and DI. The former is similar to incremental innovation (II); it carries out the benefit of enterprises on the basis of improvements of products in market. The latter is different from radical innovation (RI). Although both of them have the characteristic of disruption, DI has slower affection on current enterprises and it is also systematic (Christensen, 1997). The initial stage of disruptive technology is not a good solution relative to the existing technology, but technology can meet the need of part of the market which current enterprises neglect or think it worthless. Figure 1 shows the principle of DI.

Initially, the technology of DI is developed to meet the demands of low-end users and these demands will not attract the leading firms, but new enterprises which collect money

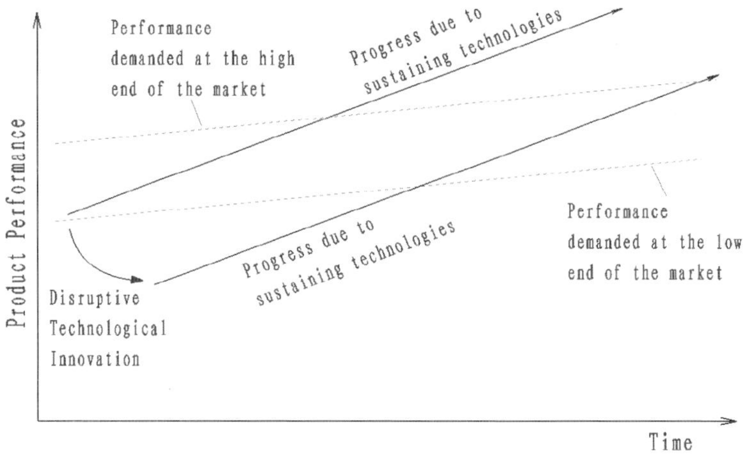

Figure 1: The impact of sustaining and DI by Clayton M. Christensen (Christensen, 1997).

within a short time and meet the needs of low ends established their companies based on the DI. The lower technology that low-end market needs may contain mass potentials. When these potentials have been developed, the new arrivals will acquire new improvements in the value chain. They will achieve the improvements made by the technology of DI as shown in Fig. 1. New techniques and the enterprises that have those techniques will become a menace to current enterprises or leading firms. As a result, the enterprises which collect money within a short time replace current enterprises, and then DI is achieved.

Innovations on technology evolution path are divided into two types in general, sustaining innovation and DI. Sustaining innovations are divided into II and RI. As shown in Fig. 2, sustaining innovation and DI constitute the whole product technology evolution path. Obviously, the DI technology can be forecasted according to the evolution line of product.

The manufacturers of most products have established a trajectory of technological improvement over time. When a trajectory of improvement has been established, those mainstream

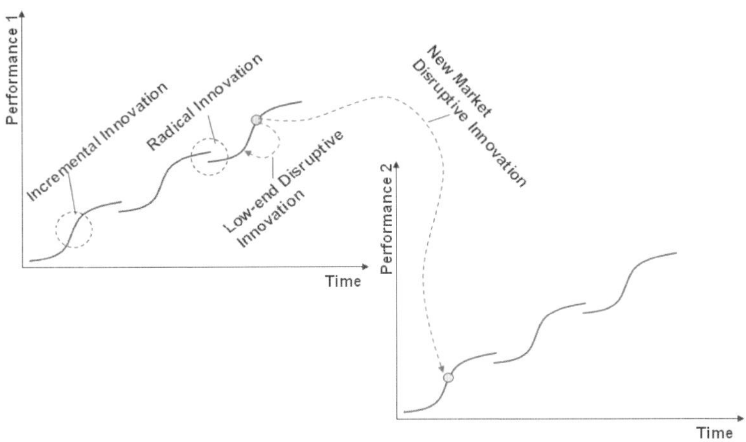

Figure 2: The DI during the process of S curve evolution.

firms occupy the markets and master advanced technology of product and there are few chances for the comer entering this market. Therefore, the new firm has to develop new product using lower technologies, current trajectory is broken.

The produce of the disruptive technology is the origin of DI, but Christensen himself is devoted to the research of commercial models of DI, market, management, and so forth. The technologies of DI in many cases described by him are already existed. His main research direction is the implements of DI based on these technologies. And he didn't do much about the technology of DI, especially the produce of this type of technology. Some scholars and firms doubt the concept of disruption too. As the cases Christensen used are the events already happened so its course is provided with chanciness or depended upon the gnosis which firms have. Therefore, the theory of DI is difficult to be used for technical forecasting of DI. Hence, in order to consummate the theory of DI, it is urgent to develop the technologies of DI in this field.

Technology is in the state of evolving. Technical evolution is affected by many elements, such as new concepts, power of the market, the status of politics, the tradition of culture, and so forth. These elements can only have an effect on the pace of technical evolution but they can't greatly change the direction of technical evolution. All the originators in the world as a whole can't be controlled or regulated. Most of their activities are separate from each other. They usually don't know other people are working over the development of technology or products, which are the same or similar to theirs and their objectives seem stochastic. However, whether a new technology or new product can be accepted by the society depends upon whether the technology or product meets the current law of technology evolution. G. S. Altshuller, who is the founder of TRIZ, and his staff found the laws of the technical evolution in different areas are same by means of the analysis of many patents. We can forecast developmental trend of technology in the future voluntarily if we predominate these laws and we can design the products for tomorrow.

The technology evolution theory in TRIZ reflects the technical systems, the composing elements, the important, steady-going, repeated reciprocity in evolution between the systems and the environment. Fry and Rivin (Fry and Rivin, 2005) reduce the technical evolution laws to nine points as follows:

(i) Law of increasing degree of ideality
(ii) Law of nonuniform evolution of subsystems
(iii) Law of increasing dynamism
(iv) Law of transition to higher level systems
(v) Law of transition to microlevels
(vi) Law of shorting of energy flow path
(vii) Law of completion
(viii) Law of increasing controllability
(ix) Law of harmonization of rhythms

Laws of technical system evolution give us directions of evolution but they didn't show us the details of each direction. There are many technical evolution paths under every law and the technologies evolution line is made up of different stations, which indicate the process and they also offered the function of technical forecasting.

The evolution path of different products is different, for example, reducing cost, increasing function, improving reliability, reducing pollution, and so forth are all the possible directions of evolution. Low cost, high function, high reliability, no pollution are the directions of evolution on condition that all products are regarded as an integrated system. The ultimate state of products evolution is called ideal final result (IFR) (Savransky, 2000). Therefore, every product is evolving toward its IFR state. Increasing the ideal level of the products ceaselessly is the objective of the innovation of the products. There are few existed products which achieve IFR. Evolving toward the direction of IFR is the development direction of almost all the products in the future.

Figure 3 shows a product technology evolution path model. The model consists of *n* states from the initial state to the top state, and these states may be classified by their position in the

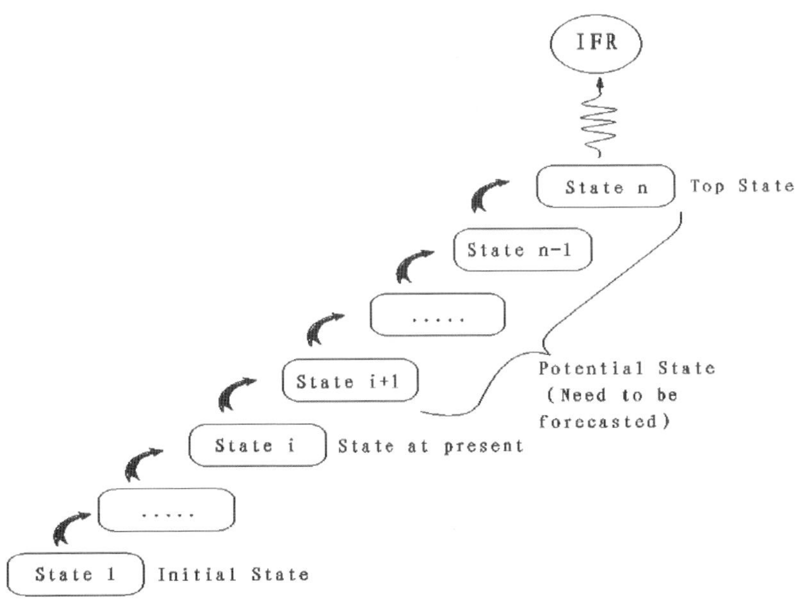

Figure 3: Technological system evolution model.

evolution path. Among them, the state *i* represents current state, and state 1 is known as initial state. From the state *i* to the state *n* is early technology evolution path, therefore, they are known technical state. The following evolution path is from the state *i* to the state *n*, which is unknown technical state needed to be forecasted.

The Bifurcation Points of Disruptive Technologies

In the whole technology evolution process of the product leading to IFR direction, the sustaining innovation and the DI are carried on in turn, where the DI divides into new-market disruptions and low-end disruptions. The former prevents the product technology system evolution unbalance caused by long-term sustaining innovation, and the latter avoids the long-term sustaining innovation leading to user's demand surplus. Just these two DI and the sustaining innovation cooperation impel the product to develop to the IFR direction.

The evolution of real systems is associated with a mix of regular, deterministic, and highly predictable events along with events that are random, stochastic, haphazard, difficult-to-predict, and so forth. A small influence often produces enormous (and unexpected) results, while in other situations extensive efforts produce an inadequate outcome (Boris and Alla, 2005). Figure 4 shows bifurcation points on the technological evolution path. A main characteristic on the bifurcation points is that their properties will change under the influence of processes taking place within them. Unfortunately, only on the evolutionary line area between two adjacent bifurcation points, technologies can be forecasted analytically, and a large number cannot be forecasted.

When a low-end DI occurs, the technological evolution appears as countermarch. As shown in Fig. 5, b to a is the process of the low-end DI.

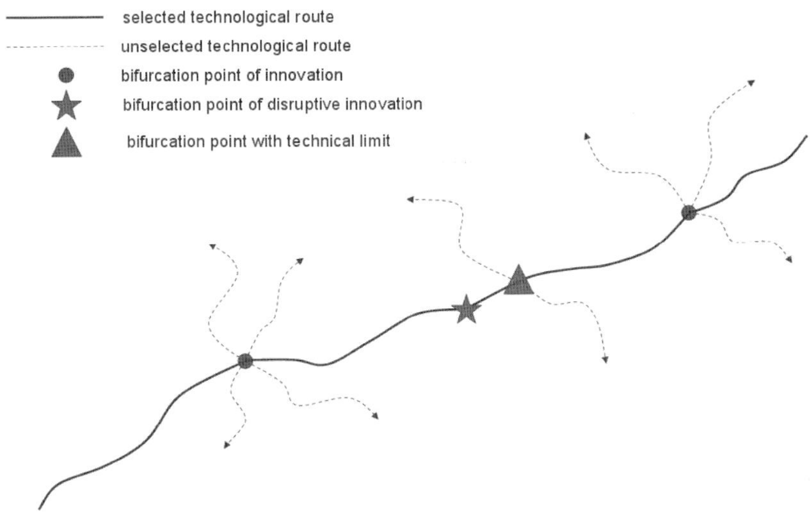

Figure 4: Bifurcation points on the technological evolution path.

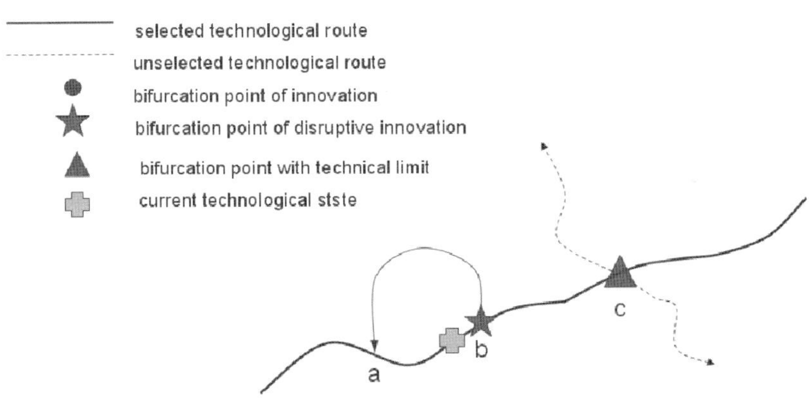

Figure 5: Bifurcation points of low-end DI on the technological evolution path.

When a new-market DI occurs, the technological evolution appears as bifurcate characteristic. As shown in Fig. 6, b to c is the process of the new-market DI.

Mixed DI is the combination of the former two DI and possesses the advantages of both of them. When a mixed DI occurs,

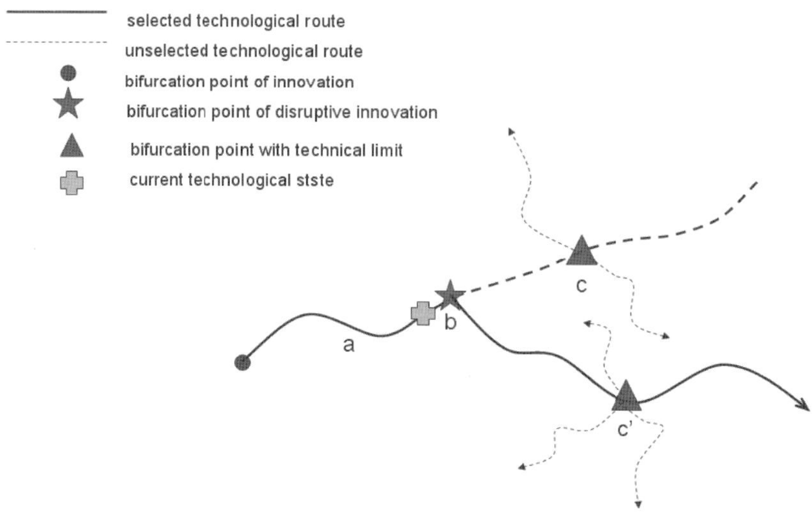

Figure 6: Bifurcation points of new-market DI on the technological evolution path.

the technological evolution appears as both countermarch and bifurcate characteristic. As shown in Fig. 7, b to a to c is the process of the mixed DI.

Composing Analysis of the Technical System for the Product

Systematology is the knowledge to research the general model, structure, and law of a system. It works over the corporate characteristic of all kinds of systems which describes the function of a system quantificationally by means of mathematical method. Systematology is seeking and establishing the principle, fundamental and mathematical model which are the same with all systems. It is a burgeoning ology which has the character of mathematics and logic. Commonly, the entia that has holistic function and synthetical action is called a system, which is made up of interrelated, interdependent, mutual checking, and interactional alternative or process (Mao, 1997).

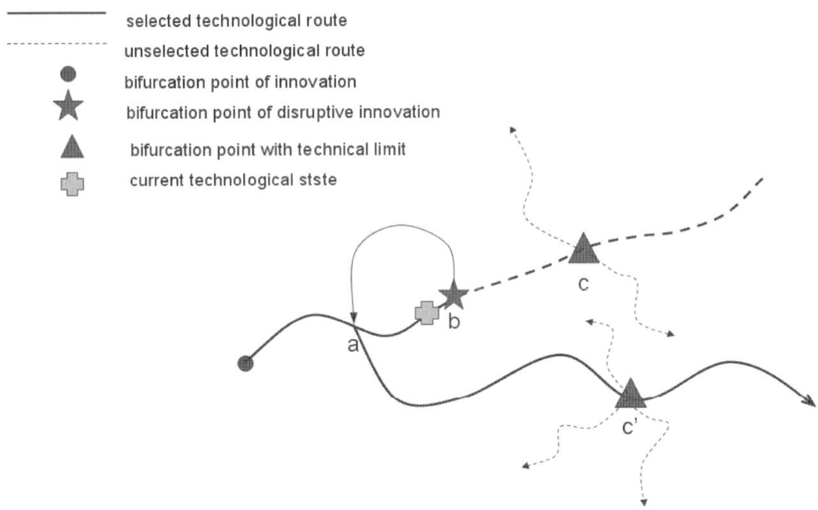

Figure 7: Bifurcation points of mixed DI on the technological evolution path.

Product is the complex entia that constitute of different components which has holistic function and synthetical performance. The technical system that forms the product is made up of subsystems at all levels and each subsystem has integrated systems structure which can be analyzed as a whole technical system. To the decomposition of a technical system which is known, we usually use tree analytic process as is showing in the picture. In order to escape the appearance of extra hierarchies of technical system decomposition, the smallest unit in the decomposition should be the outsourcing unit which is for the manufacturing of the products. Moreover, design constraint (volume and weight, price, operation accessibility, energy consumption, etc.) will be listed among all subsystems, which acts as technical subsystem for the convenience of the evolution analysis of technical system.

Figure 8 shows technical system decomposition model that face toward the products of DI. The product technical system has a hierarchical structure and is made up of basic functions

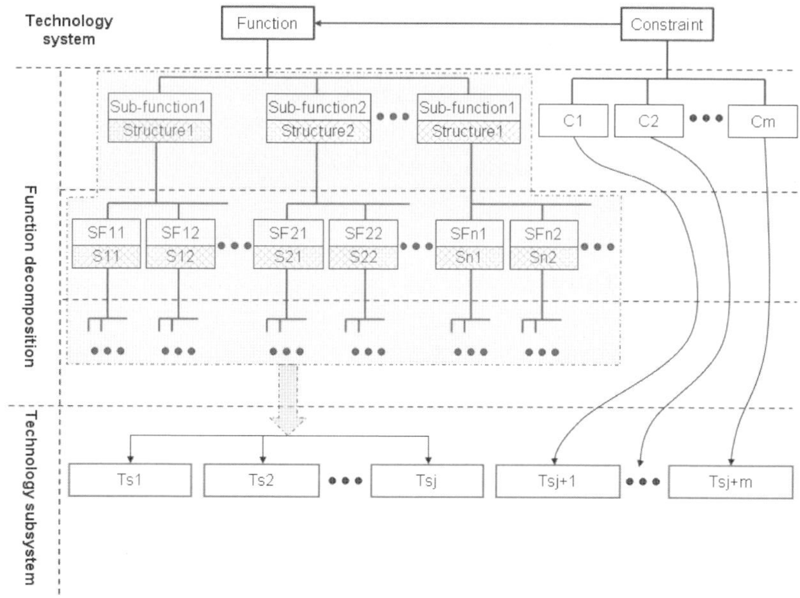

Figure 8: Model of the technology system decomposition face to DI.

and constraints, by means of a series of decomposition process; $j + m$ technical subsystems are produced at last. The decomposition process is as follows:

First, divide the basic function of the product into subfunction from technical system hierarchy, meanwhile, the constraint items which are relevant to the product should be listed as well. Please note, from the beginning of functional decomposition, the influence to the function which is brought by constraint should be considered. We may even need to add aided subfunction to meet constraint condition.

Second, all functions are decomposed with zigzagging mapping between function and structure by means of axiom design analytic process, so as to review whether the composing structure of each acquired product is the outsourcing function unit of the enterprise. Decomposition should be stopped if the composing structure of each acquired product is the outsourcing function unit of the enterprise, or else, the decomposition should continue.

Third, dividing subfunction into subfunction ceaselessly by means of homogeneous method. Then, decomposing subfunction step by step until the structure can't be decomposed any more or they are divided into outsourcing unit. Thus, the technical system decomposition tree is acquired.

Finally, the induction of technical subsystem: we may induct the front j items of the subsystem from the function tree and they unite with the mth item to become $j + m$ item, which is the final technical system. To a fixed functional decomposition tree, the concluding outcome of technical subsystem is not unique; we will bat around this point in the next section.

To the product that has highly modularization extent, the technical system decomposition is much simpler, the decomposition could be carried out according to modules, then, the design constraint can be added solely. In this way, technical subsystem is acquired. Technical system decomposition is more complex to the products whose division of components is not clear.

Evolution State of Technical System for Products

The opportunity for DI technique comes from overage evolution of product performance and the appearance of new demand market and even extends to the evolution field of technical system, which actually is the opportunity brought by the unbalancing evolution of subsystem, on condition of the restriction of user requirements. Therefore, the opportunity for DI has close relationship with evolution state of technical system.

As shown in Fig. 9, we use radar chart for expressing technical system. Supposing technical evolution system E_t is made up of six subsystems, $E_t = \{E_{t1}, E_{t2}, \cdots, E_{t6}\}$, the real line areas represent for technical performance and broken line areas represent

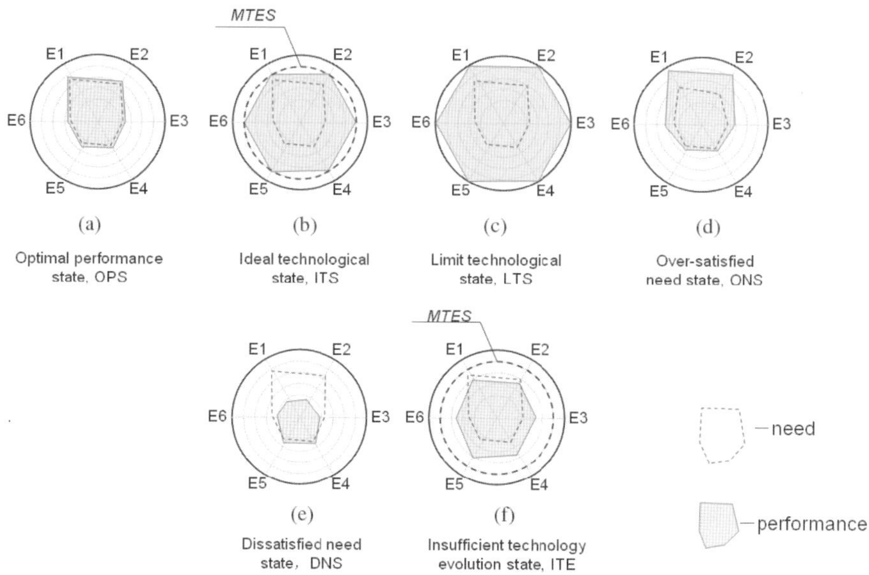

Figure 9: States of technological system evolution. (a) Optimal performance state, (b) ideal technological state, (c) limit technological state, (d) over-satisfied need state, (e) dissatisfied need state, (f) insufficient technology evolution state.

for user requirements, among others, mainstream technological evolution state (MTES) stands for the evolution state of technique which is similar with the level for social mainstream science and technology. There are six kinds of typical state in the technical system process (Sun, 2009):

(i) Optimal performance state (OPS). The performance of each subsystem can meet user requirements (Fig. 9a), at this time, systemic performance could satisfy user requirements already, moreover, the consumption of resource is at the lowest level and the price and quality of products reach a perfect balance, this is just the state which DI dreams of.
(ii) Ideal technological state (ITS). The performance of each subsystem reaches the evolution state of mainstream

technology (Fig. 9b). Technically, ITS is the most reasonable advanced state for current products, but ITS cannot come into reality in real technical system of products for the coherence of performance of technical subsystem and the cost restriction caused by cutthroat competition.

(iii) Limit technological state (LTS). The performance of each subsystem reaches its technical terminal (Fig. 9c), LTS is the object of sustainable innovation. In the process of sustainable innovation, enterprises could only improve all performance indexes at their best, which is far from user requirements, so as to win in the fierce market competition. Meanwhile, LTS cannot come into reality in real technical system of products too, as he coherence of performance of technical subsystem and the cost restriction.

(iv) Over-satisfied need state (ONS). The performance of some subsystems is obviously over user requirements (Fig. 9d). In the process of ITS, blindly improving the performance of product which is just the hotspot of market competition will lead to ONS. ONS is always regarded as the opportunity to produce low-end disruption.

(v) Dissatisfied need state (DNS). The performance of some subsystems of a product is below user requirements obviously, because the highly increasing of user requirements or the evolution deficiency of subsystems of the product (Fig. 9e). The pursuing of ITS and the inertia of technical development will lead to the lack of monitoring to user requirements. It may result in the low performance of some subsystems which is below user requirements day by day. And this is always regarded as the opportunity to produce new market disruption.

(vi) Insufficient technology evolution state (ITES). The performance of some subsystems is obviously below the evolution state of current mainstream technique (Fig. 9f). This state of subsystem has very good technical potential, the promotion of performance can be improved greatly without paying much expense.

DI Model of Technology System Evolution for Product

Technical systemic evolution appears as the resultant force of each subsystems, the process of technical systemic evolution is just the process of product innovation. Product innovation is made up of II, RI, and DI. The three innovations run through the process of technical systemic evolution.

As shown in Fig. 10, when a product is in maturity, which evolved from $S_{n-3} \Rightarrow S_{n-2} \Rightarrow S_{n-1} \Rightarrow S_n$ to current state of S_n and $E_{n1} << E_1, E_{n2} << E_2$ which we can conclude from S_n. Therefore, E_1 and E_2 are the hotspots in the market competition of the process for graduate product innovation. ONS has already appeared in the evolution process of product maturity. Moreover, user requirement to E_{n4}, which is the requirement to subsystem E_4, is always increasing. However, E_4 isn't increasing accordingly and this result in $E_{n4} >> E_4$ when the evolution

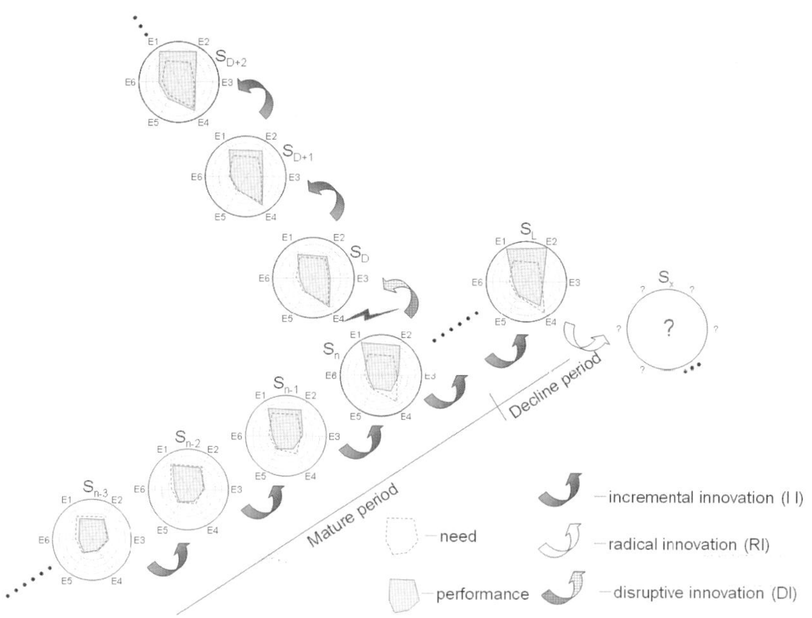

Figure 10: Technological system evolution and innovation model.

comes to current state of S_n, so DNS appears in technical systemic evolution.

S_n has two evolution potential states which are as follows, according to the definition of state S_n above:

(i) Embranchment Potential of II Evolution
The characteristic of II is going in for continuing improvement of critical performance in the market. In other words, the performances of subsystems E_1 and E_2 need to be advanced, in order to compete with other enterprises. Until the performance reaches technical high-point S_L by means of RI, systemic evolution reaches S_x which is the start point of curve S.

(ii) Embranchment Potential of DI
The unbalance of subsystem evolution will bring DI opportunity for a system, we may use two different kinds of DI strategy:

Low-End Market DI:

Decreasing the performance of subsystems E_1 and E_2, making it just meets user requirements. At this moment, in the potential state S_D, $E_{n1} \approx E_1$, $E_{n2} \approx E_2$; a part of product cost can be saved because of the knockdown of E_1 and E_2. Therefore, the price of the product could be reduced, so low-end market disruption that is for certain users can be realized.

New Market DI:

Adding the performance of subsystem E_4, which is for meeting user requirements. The adding of subsystem E_4 means the reinforcement of subsystemic function of products. Thus, products may enter to the user group who are interested in this subsystemic function and new market DI can be achieved.

The demand evolution corresponds with the technical evolution. The emergence of mainstream evolution technology is because the dominant markets are wild about the sustaining innovation, but this kind of innovation comes from a supposition on user's demand evolution. The fanaticism of sustaining innovation leads to the user's demand unbalance and surplus, which results in low-end disruption.

Figure 11 shows the flowchart of DI based on technological system evolution. First, selecting a product to perform DI according to self-condition and the situation of market. Second, technology maturity mapping by the patent analysis and market analysis. Here, we get the system's technical composition. After analyzing each subtechnology evolution path by using the TRIZ principle, we obtain the radical mainstream innovation technology and the relative lag evolution technology. In existence mainstream evolution technology, if there has the user's demand surplus phenomenon, we should identify the existence contradiction between the product technology

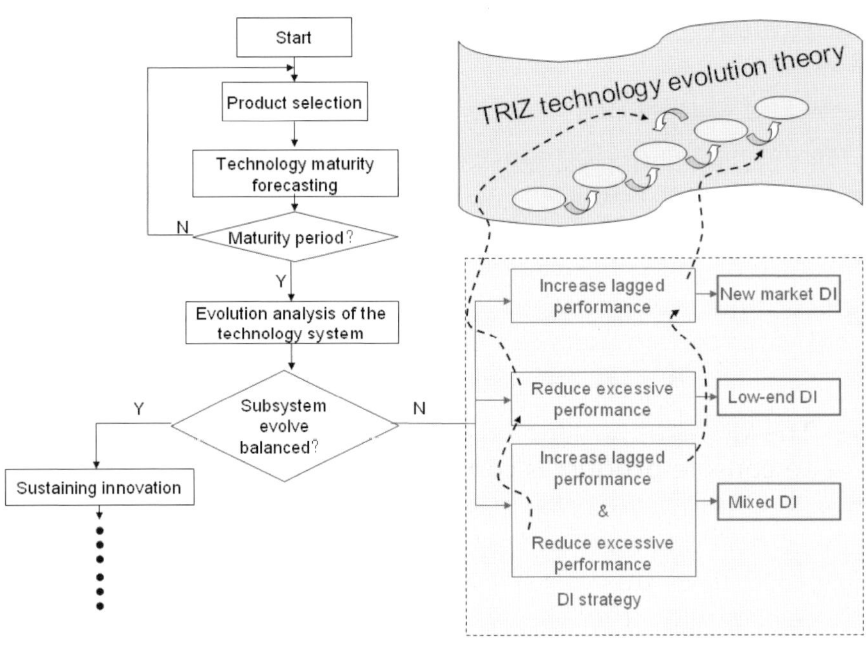

Figure 11: The flowchart of DI based on technological system evolution.

system and the mainstream evolution technology, reduce the mainstream technology rank, and strengthen the technology conflict with the product. Finally, the low-end disruption innovation product forms. In the existence lag evolution technology, we should investigate if there has such technology that it can be used conveniently for these who lack the technology and are not too wealthy. This kind of technology is the potential new-market disruptive technology. Forecasting this technology by using the TRIZ evolution path can realize the new-market DI technology.

Case Study

Video Game Technology has progressed tremendously since the dawn of its existence. The first video game system was made over three decades ago. That's a lot of years to cover. In this chapter, we discuss how video game consoles have evolved over the years.

In this final segment of our in-depth look at the evolution of game consoles, we cast our attention on the ever-present video game controller. Compared with the central processing unit (CPU) and picture quality, the game controller is easy to be disregarded. But the controller plays a great part in the game, which makes the game more fun and enjoyable. So with that, we should look into the development of game controllers, and in the process we discovered an evolution as rich and varied as the one experienced by game consoles themselves.

The first, the functional decomposition model (Fig. 12) of game console product is constructed, from that detailed function elements are obtained. By doing so, we hope to get the branched function technological evolution of the product. Regardless of the specific technique used to create a functional model, the basis identifies when an overall function is decomposed to a small subfunction and provides a common level of detail.

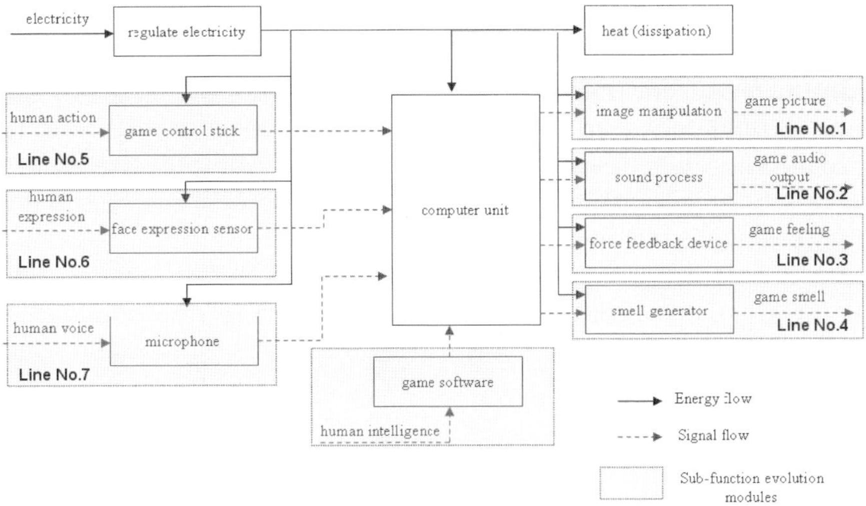

Figure 12: Function decomposition of game console.

According to the result of function decomposition, the IFR of video game console is established as shown in Fig. 13. After analyzing the video game console function in the IFR state, there are seven ideal functions in video game system, and then we can get seven technical evolution lines. According to analysis patents of video game console, every technical evolution line has been analyzed. Because of the rapid development of the picture display technology, many advanced technologies were widely used for that. From the original state till now, in order to improve the picture quality of video game, many advanced technologies had been adapted. For example, CPU has been developed from 8 bit to 128 bit and the main frequency has been improved from 4 MHz to 3.2 GHz. The audio of game system has been developed rapidly too. Compared with the improvement of the picture display technology, other features of game console such as feeling, smell, expression, and voice have little improvement all these years (Fig. 14). Consequently, its mainstream evolutionary technologies and

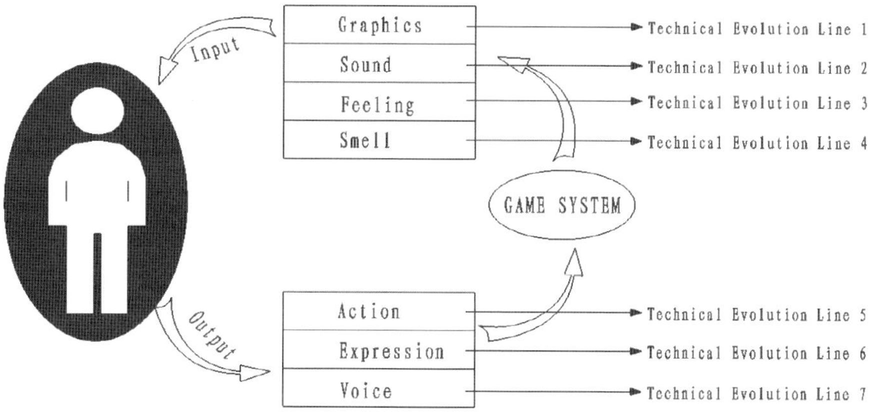

Figure 13: IFR of video game console system.

laggard evolutionary technologies were concluded (referred to Table 1).

The features of the mainstream evolutionary technologies have met the needs of the consumers, which possesses the feature of the technical demand surplus. And laggard evolutionary technological features such as game controller, which conflict with mainstream technologies, are most probably keys to the DI. We can reduce the graphics standard and improve controller standard, thus the DI is possibly achieved.

In order to achieve DI, we can reduce the graphics standard and improve the technique of controller. It is easy to reduce the graphics standard, but how to improve game controller technology is an important problem. Searching the TRIZ technology evolution line, the technical evolution of game controller cohere with the TRIZ evolution principle 4, that is to say, the technology system transition to higher level systems (shown in Fig. 15). According to the evolution line, from a separated 3D game controller to a disruptive product, Wii has been put forward in 2006 by Nintendo.

Figure 16 shows a whole evolution line of game console system. According to that, we can forecast a potential DI

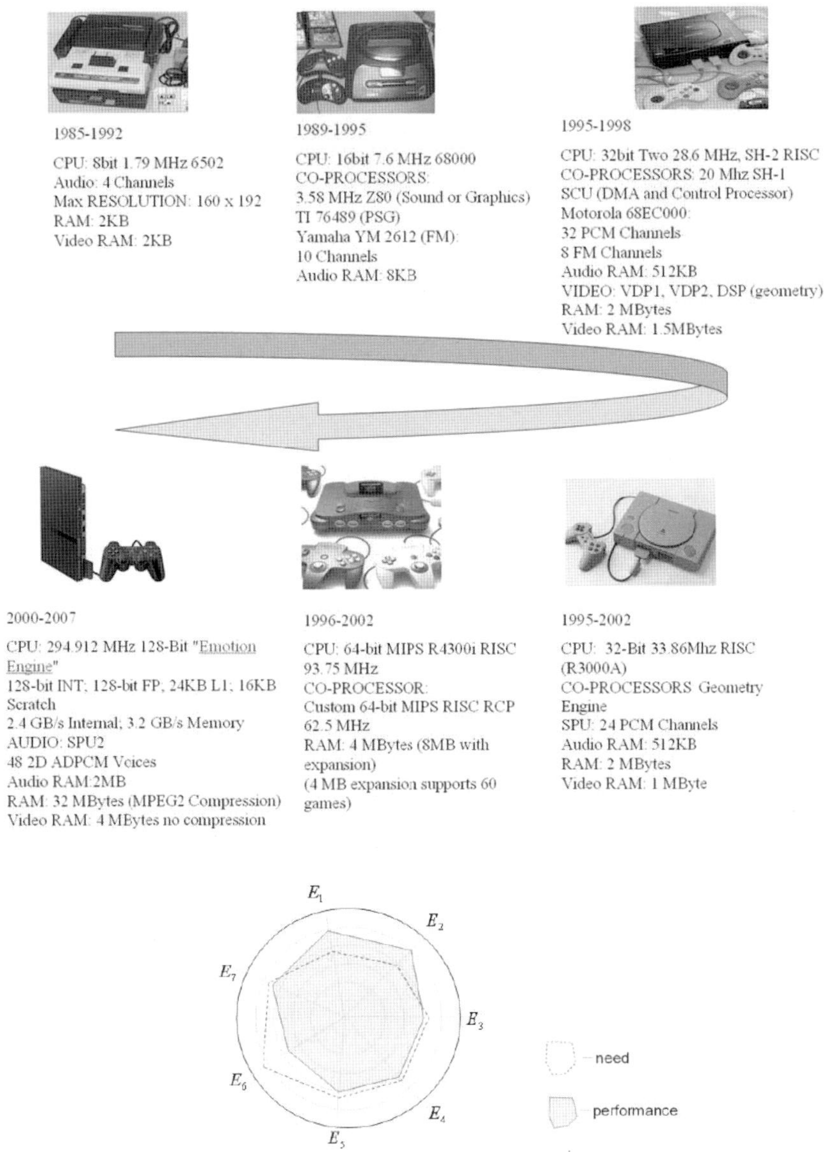

Figure 14: Technology evolution of video game console main processor.

Table 1: The classification of technical system evolution of video game system.

Classification	Subtechnology
Mainstream evolutionary technologies	Picture display technology of video game system (Line 1) Audio technology of video game system (Line 2)
Laggard evolutionary technologies	Action controller of video game console system (Line 5)

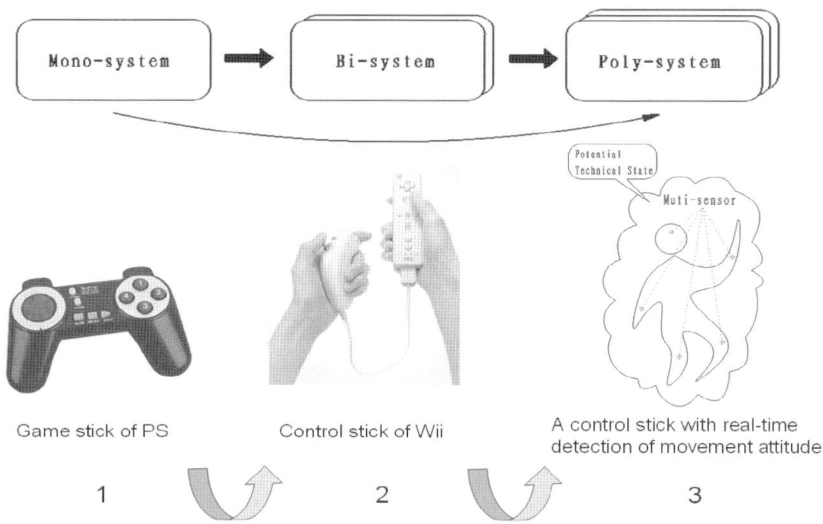

Figure 15: Forecasting game controller technology of video game console according to TRIZ evolution theory.

process: first, reduce graphics standard and improve controller standard to achieve DI and occupy the market, then improve graphics technology of game console to sustain product development in the market.

Impressively, the responsive Wii controller remains satisfying to use and player's movements can become more subtle

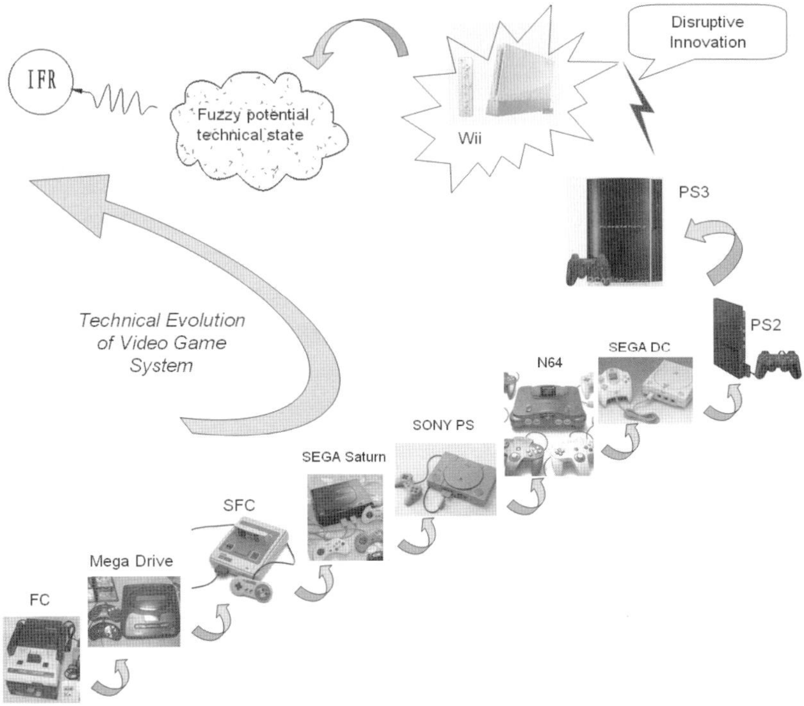

Figure 16: Whole evolution line of video game system products.

(and less energy consuming). There's also the classic controller option, and the promise of myriad forthcoming controller shells. The Wii's ridiculously enjoyable titles and innovative, motion-sensitive controllers help make it feel more like a toy you'll want to share with a group of players than a console you'd use strictly on your own for hours on end.

Because of Wii, Nintendo has officially become the most successful next-generation game console, in terms of introduction sales volume. About 600,000 units in North America helped the company to achieve a market share of about 55% in the video game console market.

To sum up, Wii is an effective DI product of game console system.

Conclusions

Based on analyzing the functional structure of product, the technology system of goal product is decomposed into subsystems, and the evolutional path of each technology subsystems is generated, respectively. In order to seek a method for searching DI technologies, the balance of technology evolution of each subsystems is studied. By analyzing each subtechnology evolution path, the radical mainstream innovation technology and the relative lag evolution technology by using the TRIZ principle, it makes technical forecasting of the DI possible. The method can be used by firms to do technical system analysis on the existing productions in the market, develop new technical market, defeat competitors, and effectively prevent their mainstream products from defeating by the DI.

Acknowledgments

This research is supported in part by the Natural Science Foundation of China under Grant Number 70972050, the Natural Science Foundation of Hebei under Grant Number E2009000070, and Foundation of the Construction of Innovation Engineer Training Base for the China Association for Science and Technology (CAST). Any opinions or findings of this work are the responsibility of the authors, and do not necessarily reflect the views of the sponsors or collaborators.

References

Altshuller, G. *The Innovation Algorithm, TRIZ, Systematic Innovation and Technical Creativity*. Worcester, England: Technical Innovation center, 1999.

Boris, Z. and Alla, Z. The Concept of Resources in TRIZ: Past, Present and Future. *Ideation International December 2004–March 2005*, Southfield, MI.

Chieh, H. C. and Dan, Y. An Exploratory Study on Understanding the Technical Dimension in DI. *The 5th International Symposium on Management of Technology*, Hangzhou, China, pp. 262–266, 2007.

Christensen, C. M. *The Innovator's Dilemma. When New Technologies Cause Great Firms to Fail*. Boston, MA: Harvard Business School Press, 1997.

Christensen, C. M. *After the Gold Rush: Patterns of Success and Failure on the Internet*. Working Papers. Belmont, MA: Innosight, 2000.

Christensen, C. M. and Bower, J. L. Customer Power, Strategic Investment, and the Failure of Leading Firms. *Strategic Management Journal*, vol. 17, no. 3, pp. 197–218, 1996.

Christensen, C. M. and Overdorf, M. Meeting the Challenge of Disruptive Change. *Harvard Business Review*, vol. 78, no. 1, pp. 67–76, 2000.

Christensen, C. M. and Raynor, M. E. *Innovator's Solution*. Boston, MA: Harvard Business School Press, 2003.

Danneels, E. Disruptive Technology Reconsidered: A Critique and Research Agenda. *Journal of Product Innovation Management*, no. 21, pp. 246–258, 2004.

Doering, D. S. and Parayre, R. *Identification and Assessment of Emerging Technologies*. New York, NY: John Wiley, pp. 75–98, 2000.

Fey, V. and Rivin, E. *Innovation on Demand*. New York, NY: Cambridge University Press, 2005.

Govindarajan, V. and Kopalle, P. K. The Usefulness of Measuring Disruptiveness of Innovations Ex Post in Making Ex Ante Predictions. *Journal of Product Innovation Management*, no. 23, pp. 12–18, 2006.

Kohlbacher, F. and Chieh, H. C. Disruptive Innovations and the Greying Market. *Proceedings of the 2007 IEEE IEEM*, pp. 1915–1919, 2007.

Kohn, S., Husig, S., and Kolyla, A. Development of an Empirical Based Categorisation Scheme for CAI Software. *1st IFIP TC-5 Working Conference on CAI*, ULM, Germany, 2005.

Mao, J. Q. *The Evolutionary Principle, Process and Model on Technological Innovation*. Beijing, China: Economy & Management Press, 1997 (in Chinese).

Rowe, G. and Wright, G. The Delphi Technique as a Forecasting Tool: Issues and Analysis. *International Journal of Forecasting*, vol. 15, no. 4, pp. 353–375, 1999.

Savransky, S. D. *Engineering of Creativity*. New York, NY: CSC Press, 2000.

Smith, R. The Disruptive Potential of Game Technologies. *Research Technology Management*, vol. 50, no. 2, pp. 57–64, 2007.

Sun, J. G. *Methodologies and Key Technologies for Roadmapping Disruptive Innovation Based on Triz*. Unpublished doctoral dissertation, Hebei University of Technology, Tianjin, China, 2009 (in Chinese).

Tan, R. H. *Theory of Innovative Problem Solving*, Beijing: Science Press, 2004.

Walsh, S. T. Road-Mapping a Disruptive Technology: A Case Study the Emerging Microsystems and Top-down Nanosystems Industry. *Technological Forecasting & Social Change*, vol. 71, pp. 161–185, 2004.

Chapter 4

Technology Assessment: A Framework for Identification of Causalities Among Internal and External Modes of Technology Acquisition

*Farshid Mohamadnejad**

Many firms gain new technologies and capabilities needed to maintain or enhance their competitiveness in today's rapid technological advancements. It is most noted that many of these companies may choose to acquire them from other firms, rather than relying on internal development or other mechanisms like alliances (Kozin and Young, 1994). Technology acquisition plays a key role in the management of technology in industrial organizations (Durrani *et al.*, 1998). It is believed that many developing countries are trying to bridge the development gap by means of technology acquisition (Castells, 1996). It is noted that a primary driver of many acquisitions resides also on the desire to obtain valuable resources, including technologies and capabilities possessed by the target firms. This motive appears to have increased in relative importance in the most recent wave of acquisition activity (Ranft and Lord, 2002). Though the number and value of acquisitions surged across a wide variety of industries during the 1990s, acquisition activity rose dramatically in high-technology sectors such as software, telecommunications,

*Corporate Design Manager, Pars Aryan Financial Holding Company, Pasargad Financial Group, Tehran, Iran.

networking, electronics, information services, and biotechnology (Sikora, 2000). The term "acquisition" refers to all the stages from buying, introducing, applying, adopting, adapting, localizing, and developing through to diffusion (Joham and Pradham, 2006). The acquisition issue is multifaceted for various reasons including large variety of organization applications, rapid change in new technology, and involvement of several business entities in the organization. The set of processes for the build, lease, or buy decision must be identical for every instance or business opportunity that arises. The processes determine the strategic value and potential savings of the proposed project, as well as factors like business transformation versus drive for competitive strategies (Chan, 2005). Well-established literatures on technology acquisition framework highlight this very fact that the technology acquisition by its nature can be regarded as a multivariate discipline which indeed has received many considerations and recognitions from different theoretical and practical research backgrounds (Wamae, 2006; Kumar *et al.*, 2004; Ahmed and Sahinidis, 2008). Then the result of this outlook is provision of a stream which enables practitioners and academics to identify and realize different kind of relationships and relationships could be existed and found among multiple perspectives and components of management of technology in organization (Ranft and Lord, 2002; Cassiman and Veugelers, 2002). Also, at the macroeconomic level of analysis, particularly in developing countries, technology acquisition plays a pivotal role to study the growth strategies pursued in the countries and to overview the market expansion or failures among them. It is argued that the most important market failures were those affecting the process of learning how to use new technologies in different sectors (Khan, 2009; Kozin and Young, 1994).

While there are ample theoretical and empirical researches on firm technology acquisition, which are focused on the study of its dependence on other constructions and components of organization, the literature pays less attention to explore and analyze the interrelatedness could be found in the main and single body of technology acquisition itself. Internal and external technology sourcing are among the main substitute modes which are embodied in the general body of firm's innovation and technology acquisition strategy. It is perceived that there is a viable complementarity can be observed between firm's own R&D, as a main component of internal technology sourcing, and external technology sourcing activities. The empirical evidences reveal the fact that many firms are deploying both strategies to accomplish and maintain their competitive advantages (Cassiman and Veugelers, 2002). The author believes that by proposing and developing a proper framework for structuring and exploring the relationships between these two main modes of technology acquisition strategy, it is possible to provide an analytical basis for better decision-making and processing at the level of managerial boards of organizations that are involving in any technology acquisition or transfer project.

Therefore, in this research, it is intended to propose and deploy decision-making trial and error laboratory (DEMATEL) (Fontela and Gabus, 1976) framework, as a powerful group decision-making model to identify and visualize structural relationships among accessible modes of internal and external technology sourcing in a firm through processing of the expertise and knowledge of organization senior managers or experts. The DEMATEL can clear see the cause and effect relationships of criteria when measuring a problem (Lin and Wu, 2004). The power of DEMATEL technique in modeling cause and effect relationships makes it possible to identify and analyze the cause and effect relationships among different modes of internal and external technology acquisition through processing of the gathered information from the senior managers or team of technology experts.

The proposed framework prepares an organizational basis for managerial boards and senior technology experts to explore the latent interrelatedness among internal and external modes of technology acquisition. And abandon the previously admitted principality, which structured internal and external modes of technology acquisition in two distinct categories, the efficiency of the proposed framework in recognizing these two modes in an integrated body and in addressing to the assessment of exploring causalities among them can be regarded as an advantage.

Internal and External Technology Acquisition: A Perspective

It is most cited that technology acquisition and transfer help create new industries, generate employment opportunities, and develop a supply network for producing advanced materials and components. Technology and innovation enable businesses to improve their processes, produce more competitive products, and make a profit (Aragon *et al.*, 2005). Therefore, it is needed continuously to explore and maintain different dimensions of technology acquisition structure in order to expand the possibilities of managing its interconnected process more efficiently.

It is believed that the firm's need for technological growth must be an important consideration when deciding among technology acquisition options. Hence, the process of internal and/or external technology acquisition sources must be recognized and analyzed in this regard. The literature shows that the performance of a firm's innovation strategy that relies on successfully integrating externally acquired technology, depends not only on a successful outcome of the innovation process based on this information, but also on the ability of the firm to appropriate the benefits from this innovation. The success of the innovation process depends on the internal research capabilities of the organization (Cassiman and Veugelers, 2000).

Internal technology acquisition is the result of technology development efforts that are initiated and controlled by the company itself. Internal acquisition requires the existence of a technological capability in the company. External technology acquisition is the process of acquiring technology developed by others for use in the company. External technology acquisition generally has the advantage of reduced cost and time to implement, and lower risks. However, almost all technology available from external sources was originally developed for different applications (Aswathappa, 2005).

Many companies rely on a technological advantage to differentiate their product from their competitors. These companies must use internal technology development via internal research and development or contracting R&D from an exclusive supplier to maintain their competitive advantage. Technology acquisitions that do not give them exclusive access to the technology make it impossible to prevent their competitors from having the same technology in their products (Aswathappa, 2005).

The literature shows that many forms of technology acquisition are a combination of external and internal activities. Combination options include the addition of networking to internal activities, reverse engineering (where internal people decipher developments accomplished by others), covert acquisition (the more blatant copying of another company's technology), contracting others to conduct R&D for you, and forming a R&D partnership where portions of the technology development are shared with others. In fact the list of options discussed in this chapter form a continuum from purely internal to purely external.

It is suggested that most of the literatures based on transaction costs concentrates on the choice between internal and external sourcing for individual transactions, as substitute modes for generating innovation (Cassiman and Veugelers, 2002). Furthermore, the appropriability regime will affect the decision to buy technology externally, as well as the mode in which the acquisition will occur. In general, it is expected that better appropriation of innovation results would lead to more technology transactions (Rahman, 2008).

It realized that firms that decide to acquire technology externally are more likely to acquire this technology in disembodied form such as through licensing agreements or R&D

contracts. Loose appropriation environments quickly erode a firm's technological advantage. In that case firms will develop specialized complementary assets internally to protect their technology (Rahman, 2008).

Given the above propositions, we can come by this conclusion that providing a process for realizing and visualizing the relationships between different modes of technology acquisition on the basis of internal or external would be an advantage.

An Overview of DEMATEL Method

The Battelle Memorial Institute conducted the DEMATEL method project through its Geneva Research Centre. The original DEMATEL method was aimed at the fragmented and antagonistic phenomena of world societies and searched for integrated solutions. In recent years, the DEMATEL method has become very popular in Japan, because it is especially practical and useful for visualizing the structure of complicated causal relationships with matrices or digraphs. The matrices or digraph portrays a contextual relation between the elements of the system, in which a numeral represents the strength of influence. Hence, the DEMATEL method can convert the relationship between the causes and effects of criteria into an intelligible structural model of the system (Lin and Wu, 2004).

Suppose a system contains a set of criteria $C = (C_1, C_2, \ldots, C_n)$, and particular pairwise relations are determined for modeling with respect to a mathematical relation.

Definition 1. The pairwise comparison scale may be designated into five levels, where the scores of 1, 2, 3, 4, and 5 represent "No influence", "Very low influence", "Low influence", "High influence", and "Very high influence", respectively. The grading scale is typically between 1 and 5 as presented in Table 1.

Definition 2. The initial direct-relation matrix Z is a $n \times n$ matrix obtained by pairwise comparisons in terms of influences

Table 1: The grading scales and their correspondent number.

Scaling	
Type	Number
No influence (No)	1
Very low influence (VL)	2
Low influence (L)	3
High influence (H)	4
Very high influence (VH)	5

and directions between criteria, in which z_{ij} is the degree to which the criterion C_i affects criterion C_j. Accordingly, all principal diagonal elements z_{ij} of matrix Z are set to zero.

$$Z = \begin{matrix} & C_1 & C_2 & \cdots & C_n \\ C_1 & 0 & z_{12} & \cdots & z_{1n} \\ C_2 & z_{21} & 0 & 0 & z_{2n} \\ \vdots & \vdots & \vdots & \ddots & \vdots \\ C_n & z_{n_1} & z_{n_2} & \cdots & 0 \end{matrix} \tag{1}$$

Definition 3. Let

$$s = \max_{1 \le i \le n} \left(\sum_{j=1}^{n} z_{ij} \right), \tag{2}$$

the normalized direct-relation matrix X can be obtained through formula (3).

$$X = \frac{Z}{s} \tag{3}$$

The DEMATEL method further assumes that at least one i such that $\sum_{j=1}^{n} z_{ij} < s$. This assumption is satisfied in almost all practical cases. Hence, matrix X just resembles the substochastic matrix obtained from an absorbing Markov chain matrix by deleting all rows and columns associated with absorbing states. It had been proved that $\lim_{k\to\infty}(X^k)=O$ and $\lim_{k\to\infty}(I+X+X^1+\ldots+X^k)=(I-X)^{-1}$, where O is the null matrix and I is the identity matrix.

Definition 4. The total relation matrix T can be acquired by formula (4).

$$T=\lim_{k\to\infty}(X+X^1+\ldots+X^k)=X(I-X)^{-1} \tag{4}$$

Definition 5. Let $t_{ij}\,(\mathrm{i,j}=1,2,\ldots,\mathrm{n})$ be the elements of the total-relation matrix T, then the sum of rows and the sum of columns, denoted as D_i and R_j, respectively, can be obtained through formulas (5) and (6).

$$D_i=\sum_{j=1}^{n}t_{ij}\quad (i=1,2,\ldots,n) \tag{5}$$

$$R_j=\sum_{i=1}^{n}t_{ij}\quad (j=1,2,\ldots,n) \tag{6}$$

Definition 6. A causal diagram can be acquired by mapping the ordered pairs of (D_k+R_k, D_k-R_k), where the horizontal axis ($D+R$), named "Prominence", is made by adding R_k to D_k, and the vertical axis (D_k-R_k), named "Relation", is made by subtracting R_k from D_k.

Following the general structure of DEMATEL technique, it is possible to omit definition 6 and replace the threshold value for completing the unfinished step. The threshold value is used to obtain the impact-digraph-map. Setting a threshold value, p, to filter the obvious effects denoted by the elements of matrix T, is necessary to explain the structure of the elements. Based on the matrix T, each element, t_{ij}, of matrix T provides information about how element i influences to element j. If all the information from matrix T converts to the impact-digraph-map, the map will be too complex to show the necessary information for decision-making (Li and Tzeng, 2009).

To obtain an appropriate impact-diagraph-map, decision-makers, who are managerial board or technology experts in our research domain, must set a threshold value for the influence level. Only some elements, here the appropriate modes of internal and external technology acquisition, whose influence level in matrix *T* higher than the threshold value, can be chosen and converted into the impact-digraph-map in the general purpose.

The threshold value is decided by the decision-maker or experts, who are managerial board or technology experts in this paper, through technical consensus discussion. Like matrix *X*, contextual relationships among different modes of technology acquisition which are located in matrix *T* can also be converted into a digraph map. If the threshold value is too low, the casualties will be too complex to show the necessary information for decision-making. If the threshold value is too high, many modes of technology acquisition will be presented as independent elements without showing the relationships with others. Each time the threshold value increases, some casualties will be removed from the modes of internal/external technology acquisition.

DEMATEL Framework as a Basis for Identification of Relationships between Internal and External Modes of Technology Acquisition

As it is mentioned in pervious sections, most of the literatures which are constructed based on transaction costs concentrate on the choice between internal and external sourcing for individual transactions, as substitute modes for generating innovation (Cassiman and Veugelers, 2002).

It is believed that companies must make their acquisition decisions with a full understanding of their internal capabilities (or lack of them). A company should not overestimate its internal capability. It should not put its people into a position of making decisions or managing situations that are beyond their knowledge base (Aswathappa, 2005). It is suggested also that although the availability of external technology may substitute for own research investment by the receiver firms, there are also arguments to stress the complementarity between in-house R&D and external know-how, as the literature suggests (Granstrand *et al.*, 1992).

Once a suitable technology is located, the firm with in-house R&D capabilities is better able to evaluate the technology. Often the external technology is only available on an exchange basis, leading to many cooperative types of sourcing. When the firm decides to buy the technology, its own R&D operations allow it to better integrate the technology because external knowledge sources do not automatically find their way into the firm's innovation process (Cassiman and Veugelers, 2002).

The multiple links between internal R&D capabilities and external technology acquisition suggest that external technology sourcing is typically embedded in the wider innovation strategy of the firm. Within this wider innovation strategy, there are also other activities that the firm might use to build up and exploit its technology base besides the traditionally considered buying of technology through licensing or R&D contracting. Compared to market transactions and internal development, R&D cooperation allows a faster, less costly and lower risk mode of accessing new technology, whereas exploiting partner complementarity and actively managing the transfers of know-how between partners (Cassiman and Veugelers, 2002).

To acknowledge the above arguments it is possible to represent and recognize an imminent principality existed in the basic substantiality of technology acquisition modes. This possibility suggests that internal and external modes of technology acquisition encompass inherent implicit interrelated relationships which possess a reciprocal characteristic. Therefore, identification and exploration of these reciprocal interrelated relationships among internal and external technology acquisition can be regarded as a basis, which empowers managerial boards to better scan the environment of their firms for management of technology efficiently.

The empirical evidences show that any exploration of relationships among the internal and external modes of technology acquisition is highly dependent on the firm's managerial board consensus. On the other hand, any process for extracting the technology acquisition casualties is a human-centric activity which could be considered as the combination and assimilation of all knowledge and preferences of the managerial boards simultaneously.

Because the DEMATEL method gathers collective knowledge to capture the causal relationships between strategic criteria and is especially practical and useful for visualizing the structure of complicated causal relationships with matrices or digraphs (Lin and Wu, 2004) we deploy it as framework to explore the cause and effect relationships among internal and external modes of technology acquisition.

For implementing the DEMATEL approach in order to identify cause and effect relationships among internal and external modes of technology acquisition, first the proceedings of management board session(s) on delineating the firm's different modes of technology acquisition must be utilized. The main result of the session(s) is the list of available modes which is acquired out of firm's technology preferences and vision.

Regarding empirical evidences (Aswathappa, 2005; Cassiman and Veugelers, 2000) we can obtain 11 modes of internal and external technology acquisition out of firm's technology management process. The selected modes are depicted in Table 2.

Following the stepwise procedure of DEMATEL framework (Li and Tzeng, 2009) brainstorming session(s) among managerial board and senior technology experts must be held to set and finalize the initial or primary cause and effect relationship among the internal and external modes of technology acquisition in

Table 2: Different modes of internal and external modes of technology acquisition.

	Modes of Technology Acquisition	Internal–External	Internal–External
1	Seizing Tacit Knowledge	X	–
2	Internal R&D	x	–
3	Internal R&D with Networking	x	–
4	Reverse Engineering	x	–
5	Technology Transfer and Absorption	x	–
6	Contract R&D	x	–
7	R&D Strategic Partnership	–	X
8	M&As	–	X
9	Licensing	–	X
10	Purchasing	–	X
11	Joint Venture	–	X

Table 2 through benefiting from the consensus among them. The result can be depicted in Fig. 1.

As it is observed clearly in Fig. 1, the extracted internal and external modes of technology acquisition contain 11 elements. So the elements of our DEMATEL framework will appear as $\textit{Technology Acquisition} = (TA_i | i = 1,2,\ldots,11)$ or as the summary $TA = (TA_1, TA_2, \ldots, TA_{11})$ for which cause and effect relationships should be extracted out using matrix transformations. As mentioned earlier, the grading scale is typically between 1 and 5 as presented in Table 1.

Next through the distributed questionnaires among 14 senior technology experts they were asked to quantify the impact of each internal and external mode over others using the grading scales of Table 1. In other words, each senior technology expert was to compare any pairs of technology acquisition mode mutually and specify the effect of each mode over the

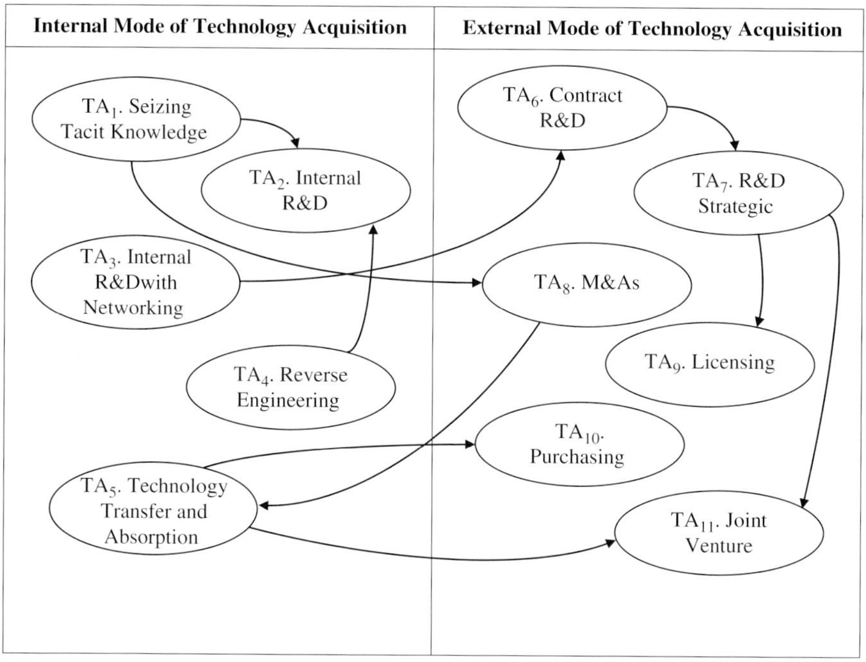

Figure 1: Initial cause and effect relationships among internal and external modes of technology acquisition.

others. As noted before, the results of this comparison form a 11×11 matrix which is the primary direct-relation matrix Z. Clearly, the number of produced matrices equates the number of senior technology experts. As an example, the viewpoints of one senior technology manager are displayed in Table 3.

Using Eqs. (2) and (3), it is possible to take the normalized direct-relation matrix out of Table 3, which is represented in Table 4. Using (4), the overall connection total-relation matrix is obtained which is displayed in Table 5.

The impact values obtained in Table 5 reveal this undeniable fact that all internal and external modes of technology acquisition have real potential to influence each other regarding integration and accumulation of the expertise and knowledge of firm's managerial boards and senior technology experts. Maybe

Table 3: A primary direct-relation matrix Z of a senior technology expert.

	TA_1	TA_2	TA_3	TA_4	TA_5	TA_6	TA_7	TA_8	TA_9	TA_{10}	TA_{11}
TA_1	—	H	No	No	No	No	No	L	No	No	No
TA_2	No	—	—	No	No	No	No	No	No	No	No
TA_3	No	No	—	No	No	H	No	No	No	No	No
TA_4	No	H	No	—	No	No	No	No	No	No	No
TA_5	No	No	No	No	—	No	No	No	No	H	VH
TA_6	No	No	No	No	No	—	VH	No	No	No	No
TA_7	No	No	No	No	No	No	—	No	L	No	VH
TA_8	No	No	No	No	VH	No	No	—	No	No	No
TA_9	No	No	No	No	No	No	No	No	—	No	No
TA_{10}	No	No	No	No	No	No	No	No	No	—	No
TA_{11}	No	No	No	No	No	No	No	No	No	No	—

Table 4: Normalized direct-relation matrix X.

	TA_1	TA_2	TA_3	TA_4	TA_5	TA_6	TA_7	TA_8	TA_9	TA_{10}	TA_{11}
TA_1	—	0.038	0.055	0.038	0.038	0.055	0.017	0.017	0.068	0.017	0.017
TA_2	0.055	—	0.068	0.068	0.055	0.068	0.068	0.055	0.055	0.055	0.055
TA_3	0.086	0.068	—	0.086	0.086	0.086	0.086	0.051	0.051	0.051	0.051
TA_4	0.068	0.086	0.068	—	0.086	0.086	0.086	0.051	0.051	0.086	0.086
TA_5	0.086	0.086	0.086	0.086	—	0.068	0.025	0.031	0.025	0.025	0.031
TA_6	0.086	0.051	0.086	0.086	0.086	—	0.086	0.031	0.025	0.025	0.025
TA_7	0.086	0.086	0.086	0.086	0.068	0.068	—	0.086	0.051	0.031	0.068
TA_8	0.025	0.021	0.025	0.021	0.025	0.025	0.044	—	0.031	0.042	0.042
TA_9	0.038	0.038	0.051	0.051	0.038	0.038	0.051	0.051	—	0.051	0.038
TA_{10}	0.031	0.031	0.031	0.038	0.038	0.038	0.031	0.031	0.038	—	0.031
TA_{11}	0.086	0.086	0.086	0.086	0.086	0.086	0.086	0.086	0.051	0.051	—

Table 5: Total-relation matrix *T*.

	TA_1	TA_2	TA_3	TA_4	TA_5	TA_6	TA_7	TA_8	TA_9	TA_{10}	TA_{11}
TA_1	—	0.062	0.038	0.038	0.027	0.028	0.044	0.056	0.041	0.026	0.017
TA_2	0.042	—	0.085	0.041	0.026	0.018	0.016	0.039	0.027	0.028	0.038
TA_3	0.046	0.044	—	0.018	0.012	0.085	0.044	0.012	0.010	0.018	0.031
TA_4	0.044	0.064	0.033	—	0.085	0.046	0.038	0.022	0.038	0.016	0.038
TA_5	0.038	0.044	0.038	0.038	—	0.031	0.026	0.028	0.011	0.056	0.085
TA_6	0.044	0.085	0.042	0.048	0.029	—	0.085	0.016	0.018	0.016	0.021
TA_7	0.038	0.044	0.048	0.026	0.068	0.028	—	0.016	0.056	0.028	0.014
TA_8	0.028	0.026	0.028	0.021	0.056	0.018	0.044	—	0.016	0.042	0.028
TA_9	0.021	0.028	0.028	0.018	0.019	0.028	0.017	0.038	—	0.044	0.038
TA_{10}	0.038	0.038	0.018	0.021	0.038	0.044	0.044	0.020	0.018	—	0.029
TA_{11}	0.034	0.036	0.039	0.018	0.044	0.040	0.038	0.016	0.018	0.028	—

this real observation can best reveal the true complexity yet intricacy of empirical casualty, which existed in an inherent nature of internal and external modes of technology acquisition.

However, it must be noted certainly that this realization is not necessarily useful and somehow overshadows the main purpose in implementing any mode of internal and external technology acquisition. In order to bestow effectiveness to the DEMATEL framework in measuring cause and effect relationships and simplify the underlying complexity in internal and external modes of technology acquisition a threshold value is defined (Li and Tzeng, 2009). Tuned to the digits of Table 5 which is actually the tabular diction of the overall connection matrix, 0.05 is defined as the threshold value so that impact values less than 0.05 will not ground for establishing relations among the selected modes.

By considering threshold values higher than 0.05, the cause and effect relationships among internal and external modes of technology acquisition can be realized. In other words, the relationships whose produced threshold values are higher than 0.05 are accepted as cause and effect relationships and other integrations are rejected out from the desk. The final result is displayed in Fig. 2.

The illustrated result in Fig. 2 shows specifically that the proposed framework has maintained the presupposed relationships among the internal and external modes of technology acquisition, besides explored, and identified three latent relationships which the senior technology experts have neglected in their initial judgmental analysis. This is one of the main advantages of the proposed framework which can be regarded as a powerful basis for empowering the managerial boards for more efficiently decision-making and analysis.

Conclusion and Remarks

Using this procedure, it is possible to introduce an integrated structure for technology acquisition process in a macro-level

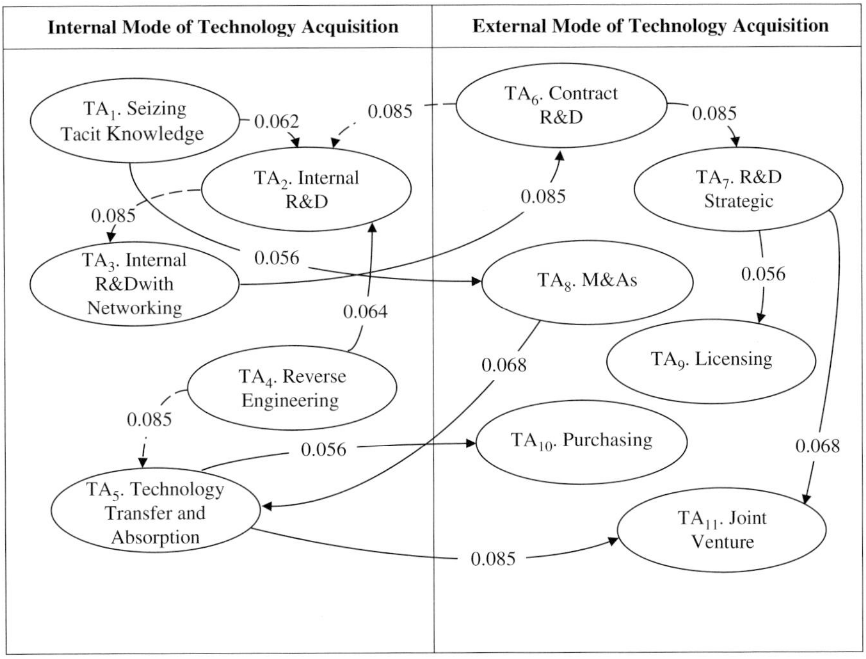

Figure 2: Final cause and effect relationships among internal and external modes of technology acquisition.

perspective. This macro-level integrated framework can be utilized and developed as a decision support system which can mathematically identify and explore the casualties among the internal and external modes of technology acquisition based on processing the collective knowledge and expertise of firms managerial boards and senior technology experts.

For future research outlines, we suggest other structural modeling approaches be engaged for identification and visualization of cause and effect relationships and their efficiency and profitability be compared relative to those of the DEMATEL approach.

Because the assigned preferences between the internal and external modes of technology acquisition are not crisp necessarily, and managerial boards and senior technology experts' domain knowledge could be extracted in a fuzzy environment, then it is suggested to extend and deploy fuzzy DEMATEL is proposed to deal with the ambiguities inherent of such the judgments.

References

Ahmed, S. and Sahinidis, N. V. (2008). Selection, acquisition, and allocation of manufacturing technology in a multi-period environment, *European Journal of Operational Research*, **189**(3), 807–821.

Aragon, C. T., Uy-Anacleto, M. R. B. and Matundan, L. P. (2005). Technology acquisition and transfer, *CACCI Journal*, **2**: 1–13.

Aswathappa, K. (2005). *International Business*, Second Ed. New York: McGraw-Hill.

Cassiman, B. and Veugelers, R. (2002). Complementarily in the innovation strategy: Internal R&D, external technology acquisition, and cooperation in R&D, C.E.P.R. Discussion Papers, 3284, London.

Cassiman, B. and Veugelers, R. (2000). *External Technology Sources: Embodied or Disembodied Technology Acquisition*. Barcelona: University Pompeu Fabra, Economics and Business Working Paper, 444.

Castells, M. (1996). *The Rise of the Network Society*. Oxford: Blackwell.

Chan, C. S. (2005). Build, buy, or subscribe? Computer World, Computer World. Retrieved Nov 26, 2006, from http://www.cw.com.hk/computerworldhk/article/articleDetail.jsp?id=153871.

Durrani, T. S., Forbes, S. M., Broadfoot, C. and Carrie, A. S. (1998). Managing the technology acquisition process, *Technovation*, **18**: 8–9, 523–528.

Fontela, E. and Gabus, A. (1976). *The DEMATEL Observer*. Geneva: Battelle Institute, Geneva Research Center.

Joham, C. and Pradham, J. (2006). IT acquisition: The need for a systematic IT strategy, Retrieved Nov 26, 2006, from http://www.ifip.org/.../openbiblio/opac/viewDocument.php?id=48&PHPSESSID=ff38181250ee3a1270b92c675988965f.

Granstrand, O., Bohlin, E., Oskarsson, C. and Sjöberg, N. (1992). External technology acquisition in large multitechnology corporations, *R&D Management*, **22**: 111–133.

Khan, M. H. (2009). *Learning, technology acquisition and governance challenges in developing countries*, Technical Report. London: School of Oriental and African Studies.

Kozin, M. D. and Young, K. C. (1994). *Using Acquisitions to Buy and Hone Core Competencies*. New York: Mergers & Acquisitions, 21–26.

Kumar, R., Krishnamoorthy, M. and Cardenez, M. A. (2004). Technology acquisition and innovation: An integrative approach for organizational success, Second World Congress on POM, Cancun, Mexico.

Li, C.-H. and Tzeng, G.-H., 2009. Identification of a threshold value for the DEMATEL method using the maximum mean de-entropy algorithm to find critical services provided by a semiconductor intellectual property mall, *Expert Systems with Applications*, **36**(6): 9891–9898.

Lin, C.-L. and Wu, W.-W. (2004). A fuzzy extension of the DEMATEL method for group decision making, *European Journal of Operational Research*, **156**: 445–455.

Rahman, A. A. (2008). Buyer-supplier relationships in advanced manufacturing technology acquisition and implementation in Malaysia, *International Journal of Economics and Management*, **2**(1): 95–126.

Ranft, A. L. and Lord, M. D. (2002). Acquiring new technologies and capabilities: A grounded model of acquisition implementation, *Organization Science*, **13**(4) 420–441.

Sikora, M. (2000). M&A Almanac: 1999 M&A Profile, *Mergers & Acquisitions*, **35**: 25.

Wamae, W. (2006). *A Technology Acquisition Model: The Role of Learning and Innovation*, Working Paper Series. Tokyo: United Nations University.

CHAPTER 5

Technology Assessment: Technological Capability Assessment for Automotive Parts Manufacturers

Mehdi Mohammadi, Mahdi Elyasi,† & Mostafa Mohseni Kiasari**

Today, technology plays a crucial role in firms' competitiveness. Therefore, it needs, like other sources of competition, to be managed from a strategic point of view (Arasti and Pakniat, 2006). Strategic view of technology should consider different aspects of technological capability (TC) of the firm, and therefore assessing the TC will lead to better strategic planning and technology development. In the manufacturing industries, impact of technology as a source of competitive advantage is widely accepted and developed (Phaal *et al.*, 2001, 2004). VanWyk (2004) pointed out that: "We need a simple, comprehensive grasp of technology. We need to understand at the macro level how technology is composed, how it evolves and how it interacts with other systems". However, its impact is complex and often difficult to be integrated into strategic management process. There is now a broad consensus among different academics, industrial experts, and policy makers that a manufacturing organization to be competitive in the global market, must be able to produce required high-quality products

*Faculty of New science and Technologies University of Tehran Gargar Shomali (Amirabad) St., Tehran, Iran.
†Department of Management and Accounting Allameh Tabataba'i University, Nezami Ganjavi Ave. Tavanir Ave, Tehran, Iran.

being reliable and economic (efficient) by easiest ways of production (Shamsuddin and Bititci, 2006).

Automobile industry is one of the most important manufacturing industries in Iran, and it has very important role in national economy. A very important element in this industry is automobile parts suppliers. Therefore, focusing on their capabilities and developing strategies to improve them is very important in Iran economy.

Technological aspects can be added to other aspects of competitiveness such as quality, cost, and financial elements through technological capability assessment (TCA) of company in this industry. In this work, to assess these technological capabilities in automobile parts suppliers in Iran and to help improve management of technology a model will be suggested. TCA must be incorporated into the overall performance indicators of a firm. Performance measurement systems (PMSs) allow mindful decisions to be made and actions to be taken as because it quantifies the efficiency and effectiveness of the past activity (operation) through the acquisition, collation, sorting, analysis, and interpretation of appropriate data. Incorporating technological competency measures into the PMS helps decision makers adopt, adapt, absorb, and utilize the appropriate technologies. From the technology management perspective, different activities such as identification, assessment, selection, acquisition, utilization, and protection of technology are very important (Phaal, 2001), therefore TCA will lead to better management of technology. In addition, the assessment process improves quality and expands existing and potentially new technology bases of a company (Panda and Ramanthan, 1996). The assessment also benchmarks and identifies the strengths and weaknesses of a company (Panda and Ramanthan, 1996), and will improve the capability of strategic technology planning and technology development in a company.

Supplier of Automotive parts company (SAPCO) is the biggest automotive parts supplier in Iran and many of Small and Medium Enterprises (SME) have close relationship with it.

In this chapter, 10 automotive parts manufacturers, which are under the superintendence of SAPCO, have been analyzed.

In this chapter, we first tried to define the technology and examine different aspects of its nature. After studying different models of TCA and their efficiencies, a number of these models were selected as the foundation of assessment model. However, considering that utter use of these models and their proposed indices have no effective application (are not effectively applicable) in Iran, so localized model was formulated, and then, using questionnaire, automotive parts manufacturers were appraised with regard to capability level.

The Concept of TCA

What is technology? What is its origin and its nature? To assess, what aspects of technology should be considered? These are the fundamental questions about technology and its capability. In this section, the focus will be on the answers to the above questions.

Technology has been defined in several ways. Some define technology as a process of converting input into output with the aim of facilities and methods (porter, 1985). According to Khalil (2000), "Technology can be defined as all the knowledge, products, processes, tools, methods, and systems employed in the creation of goods or in providing services". He continues "only when knowledge is practically implemented to create new things, operate a system, or provide a service that we enter the realm of technology". But the most important aspect of the technology is the capability. We must consider that technology should be seen as a created capability.

In a classification, technology has been used with different meanings:

- USAGE 1: HARDWARE (OR ARTIFACTS): Possible denotation: nonnatural objects, of all kinds, manufactured by humans.
- USAGE 2: SOCIOTECHNICAL SYSTEM OF MANUFACTURE. Possible denotation: all the elements needed to manufacture

a particular kind of hardware, the complete working system including its inputs: people; machinery; resources; processes; and legal, economic, political, and physical environment.
- USAGE 3: The information, skills, processes, and procedures for accomplishing tasks: Possible denotation: KNOWLEDGE, TECHNIQUE, KNOW-HOW, OR METHODOLOGY in the usual sense of these words.
- USAGE 4: A SOCIOTECHNICAL SYSTEM OF USE is a system using combinations of hardware and people (and usually — other elements) to accomplish tasks that humans cannot perform unaided by such systems — to extend human capacities (Kline, 2003).

In the Asian and Pacific Center for Transfer of Technology (APCTT 1989) definition of technology, there are four elements:

- Techno ware: hard aspect of technology like facilities and machines.
- Info ware: soft aspect of technology showing the information and explicit knowledge of technology.
- Human ware: soft aspect of technology showing the tacit knowledge in human activities.
- Orga ware: soft aspect of technology showing the organizational aspect of technology.

Therefore, technology is composed of hard and soft aspects. Facilities, skills, and knowledge are the most important elements of technology (Zeleny, 1986). The nature of technology is not fully explicit, but rather it has important tacit facets. The nature of technology has two aspects; its first aspect which is explicit in essence and is manifested in the form of information can be effortlessly transferred from place to place; and the second one is tacit taking the form of firm-specific knowledge which can easily not be moved from a place to another one (Radosevic, 1999; Johnson *et al.*, 2002; Nonaka and Takeuchi, 2004).

Our proposed model has incorporated both explicit and tacit aspects when assessing technology capability leading to an important result. Based on these arguments different approaches of TCA were studied and analyzed to see how they address the tacit and explicit elements of technology. In the next section, we address some important aspects of different approaches in TCA.

Technological Capability

Over the past decade, firms' TC has been an important strategic resource for them to achieve competitive advantage within their industry, particularly in high-tech industries (Duysters and Hagedoorn, 2000). TC has been defined by several researchers and is often used in different contexts. The definition of TC is varied in perspective, depending on the aims of the researchers (Zedtwitz and Jin, 2004). Some define TC from strategic perspective and others define it in practical sense. TC is the ability to perform required technical activities within the firm, such as capacity to develop new products and processes and effective operation of facilities Technological capabilities can be divided into three major categories. The first one, strategic technological capabilities, includes creation, design and engineering, and construction capabilities. The second, tactical technological capabilities, encompasses all functional capabilities such as production, marketing and selling, and servicing capabilities and the third, supplementary technological capabilities, entails acquiring and supportive (training; planning; information support and networking; technology selling; and safety and security) capabilities (Panda and Ramanathan, 1997). Lall (1990) defined TC as the ability to perform all the technical functions involved in operating, improving, and modernizing the firm's facilities and to make them productive. From a different point of view, TC is the ability of the company to effectively capitalize on the technology management knowledge (Janeš, 2007). Zedtwitz and Jin (2004) defines TC as the capability to effectively explore the technical knowledge and skills, in effort to improve and develop new products, processes, and existing technologies and to create new knowledge and skills so as to meet the challenges of increasingly competitive environment. Štrukelj and Dolinšek (2011) refer to TC as our capacity to utilize technologies to achieve our objectives in an effective and successful way. He continues "Technological capability of companies refers to competence/capacity of companies to use technologies as well as knowledge and skills necessary for their proper use in a way that consequentially guarantees value maximization and profits for investors". Thus, TC enables firms to promote their innovation and improve their products and processes in an industry characterized by high levels of dynamism and rivalry (Ortega, 2010) and it is one of the most important sources of sustainable competitive advantage (Coombs and Bierly, 2001).

Technology Capability Assessment Approaches

There are diverse approaches for TCA and in literature it is called technology auditing. These two terms are interchangeably used in this chapter. Some of these approaches are broad and encompass all aspects of technology in the organization, but some others are narrow regarding technology in the organization. Table 1 classifies diverse methods and models briefly.

Simply defined, technology capability assessment or technology auditing evaluates the internal technology status of an enterprise and compares it with the state-of-the-art examples in the world. It then matches the management capabilities of the enterprise with its technological standing.

APCTT (1989) has a way of assessing technology capability with four dimensions that include techno ware, info ware, human ware, and orga ware. Through this way, first the complexity of each dimension is assessed and then compared with state of the art of technology. It is a quantitative approach which considers all aspects of technology in an organization.

Table 1: Technology auditing models.

Model or Method	Explanation	Source
Value chain of activities	In order to assess the technological capabilities, activities in which these abilities are concealed will be evaluated. These activities will be compared with similar challenging activities.	Porter (1985)
Panda and Ramanathan method	Four sets of criteria are introduced for carrying out evaluation and standards are introduced for each. Then firms are compared based on these standards.	Panda and Ramanathan (1996)
Direct assessment of technological capability	Levels are introduced for technological capabilities which the author calls positions. The position of each firm is compared with the position of competitors or the ideal situation.	Arasti (2004)
Atlas	Based on the four perspectives of technology components, the technological capability of one firm or one country is evaluated. Based on these four perspectives, the index is defined for the technological capabilities of the system in mind.	APCTT (1989)
The technology management audit	Includes a set of guidelines for identifying and selecting technological opportunities and analyzing causes of technological gaps in the firm.	Lindsay (1999)

(*Continued*)

Table 1: (*Continued*)

Model or Method	Explanation	Source
Technology management process assessment	By identifying important occupation unit, technology management activities are identified in a certain framework. Afterwards, these activities are assessed.	Pall (2001)
Technology audit model (TAM)	This model, which includes a set of standards for technology audit, attempts at analyzing the company's position in technology management.	Khalil (2000)
Lin method	This method proceeds at analyzing technology management capability with focus on technology acquisition through transfer. On this basis, six sets of criteria are identified and indexes are defined for each. Then, based on these indexes, the firm's capability in transfer of technology is evaluated.	Lin (1997)
Technological capability audit model[1]	This model proposes a principle of TC audit in individual organizations. According to this model, first, appropriate model of TC should be developed, then assessment tools and methods are required, data analysis and possible problems and failures identification will come next, finally, based on an appropriate model of TC and audit results, possible solutions have to be developed.	Štrukelj and Dolinšek (2011)

Source: Bagheri Moghaddam *et al*. (2011).

[1] The "Technological capability audit model" has not been in original table and was added by the authors.

Phaal *et al.* (2001) developed a methodology for technology management assessment. They focused on the main functions of technology management and developed a model to assess different levels of technology management in the organization. Their assessment approach consists of three main stages:

- The strategic overview defines a framework for linking technology with business objectives and enables selection of areas for more detailed appraisal.
- The process overview focuses on the business technology area selected in stage (i), mapping and assessing technology management leading to the identification of specific processes for more detailed assessment.
- The process investigation focuses on mapping and assessment of specific process areas during stage (ii).

The aim of this approach is to undertake a structured/systematic evaluation of a firm's technology management practices and to identify areas in need of improvement. This qualitative approach instead of focusing on the content of technology zeros in on the functions of technology management and its strengths and weaknesses in the development of the key technologies of organization.

Another approach akin to the Phaal (2001) approach is the Ford (1996) approach. In this approach, technology capability of an organization is similar to technology management capabilities in selection, acquisition, utilization, development, and diffusion of technologies. This approach is thoroughly qualitative and, regardless of the content of technology, only focuses on the functions of technology management (Ford, 1996).

Panda and Ramanthan (1996) suggest a five-step model in conducting TCA. The steps are as follows:

- Identification of value adding stages
- Identification of the TC needed to perform the necessary value addition

- Development of a set of indicators for assessing each TC
- Benchmarking of TC of the firm after a state-of-the-art company, finding the existing level of TC
- Finding of the gaps in TC (Panda and Ramanthan, 1996)

This model is based on the evaluation of organizational capability in value-added creation and through these major criteria the technology is assessed. It is both qualitative and quantitative approach taking into account all aspects of technology in an organization. In this model, both the tacit and explicit aspects of technology are analyzed.

Porter (1985), considering organizational value chain, proposed some implications for technology assessment in organization. Technologies are divided into two categories with respect to organizational value chain: core technologies and support technologies. In this model, based on the process approach, assessment of technologies will lead to performance assessment of the whole organizational value chain. Although this qualitatively considers all aspects of technology in an organization, it overlooks the content and complexity of technology and just focuses on its output and performance. In this model, just the explicit aspect of technology is analyzed and tacit aspect is not counted.

Chiesa (2001) takes into consideration the technology assessment and selection in the process of technology strategy and research and development (R&D) strategy. This approach results in the prioritization of technologies and provides a condition for technology selection and development. It is highly qualitative and is not an inclusive approach for TCA.

The important steps in the technology assessment are as follows:

- Identify the current technologies in the various sublevels.
- Characterize each technology in terms of people, processes, and system requirements.

- Map the different technologies in the framework according to the characteristics of the fundamental functions, life cycle, and hierarchy.
- Do a projection of the technology map onto the process map; this projection indicates which technologies are empowered by which processes.
- Quantify the impact of the technology on the company, using various analytical techniques (Shamsuddin and Bititci, 2006).

It is both qualitative and quantitative approach for TCA and it has a broad approach regarding assessment of technology in an organization.

With respect to these approaches, we can design an integrated model based on holistic approach to TCA in organization and consider both qualitative and quantitative indicators while analyzing both the tacit and explicit aspects of technology.

Developing a Conceptual Model for TCA of Automotive Parts Manufacturers in Iran

In this section, we propose a model for TCA of automotive parts manufacturers in Iran. In this model, a holistic approach to TCA in organization is taken and both qualitative and quantitative indicators are included when analyzing both tacit and explicit aspects of technology.

Our overall conceptual model is shown in Fig. 1. Based on different approaches reviewed in the previous section, we shaped different dimension of our conceptual model.

Like APCTT (1989), Panda and Ramanthan (1996), Ford (1996), and Phaal (2001), we believe that organization, strategy, and process of technology management must be considered in our conceptual model. Therefore, the first dimension of our model is *organizing and managing TC*. This dimension

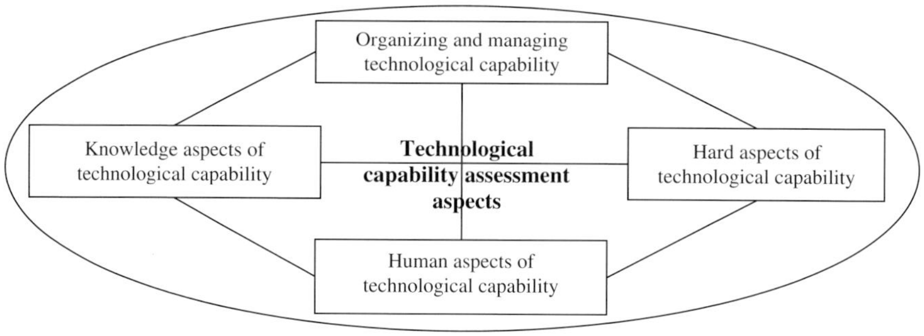

Figure 1: Conceptual model for technological capability assessment aspects.

is characterized by more tacit knowledge and less explicit knowledge.

Based on technology definitions and some approaches suggested by APCTT (1989), Porter (1985), Panda and Ramanthan (1996), Chiesa (2001), and Wet (2002), we concluded that hard aspects of TC like machines and facilities are important. Therefore, the second dimension of our model is *hard aspects of TC*. This dimension is characterized by more explicit knowledge and less tacit knowledge.

In relation to technology definitions that encompass human aspects of technology such as skills and tacit knowledge rooted in human capabilities, and based on some approaches like APCTT (1989) and Panda and Ramanthan (1996) we concluded that human aspects of TC, must consider in our model. Therefore, the third dimension of our model is *human aspects of TC*. This dimension is characterized by more tacit knowledge and less explicit knowledge.

Finally, based on technology definitions that encompass the information and knowledge aspects of technology and with respect to some approaches like Porter (1985), APCTT (1989), and Panda and Ramanthan (1996) we concluded that *information and knowledge aspects of TC* should be considered in our model. Therefore, the fourth dimension of our model is information and knowledge aspects of TC. This dimension is characterized by more explicit knowledge and less tacit knowledge.

To develop our indicators for TCA, we draw two axes in the model to illustrate them. These two axes are presented in Fig. 2 and are defined as follows:

- First axis depicts quantitative or qualitative indicators.
- Second axis exhibits tacit and explicit aspects of technology. It helps us assess these aspects of technology in the model which results in better indicators for their assessment.

In other words, to develop a holistic model, we should consider both tacit and explicit aspects and on the other hand both qualitative and quantitative aspects of technology. So, there are two orthogonal axes, which one is a spectral of qualitative and quantitative aspects of technology and another is a spectral of

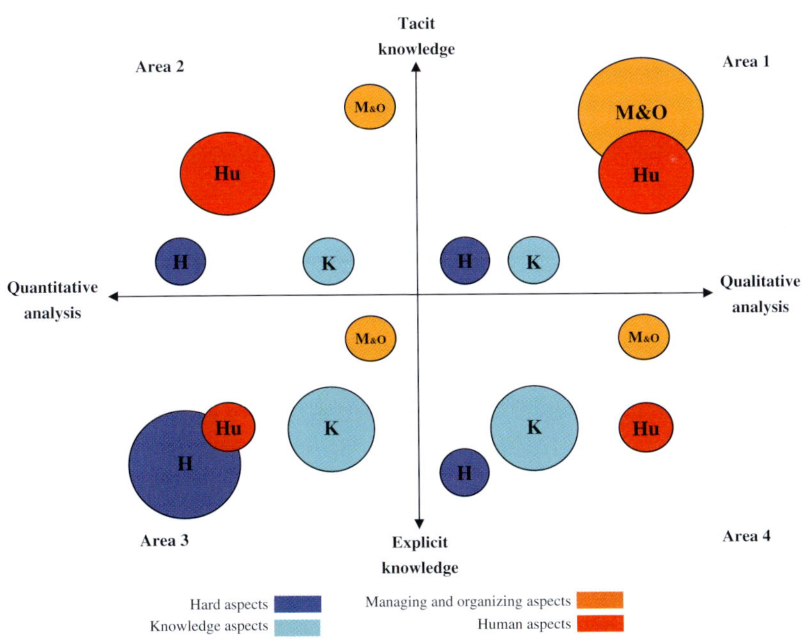

Figure 2: Taxonomy of indicators for technology capability assessment in our model.

tacit and explicit aspects of technology. And, as a result four areas are formed.

Based on this categorization, the model includes all kinds of indicators in the TCA of organization. The first category of indicators, located in area one, is related to qualitative assessment of tacit knowledge. The second category, located in area two, is related to quantitative assessment of tacit knowledge. The third, located in area three, is related to quantitative assessment of explicit knowledge. Finally fourth category, located in area four, is about qualitative assessment of explicit knowledge.

Then position of all aspects of TCA should be shown in each of these areas. In fact, some aspects of TCA are better related to some areas, so we should draw bigger circles to show this relatedness. Based on our model, managerial and human aspects are more related to assessment of tacit aspects of technology, and knowledge and hard aspects are more related to explicit assessment of technology. Moreover, managerial and knowledge aspects are more related to assessment of qualitative aspect of technology, and hard and human aspects are more related to assessment of quantitative aspect technology. Therefore, as shown in Fig. 2, our conceptual model encompasses composite indicators of TCA ranging from qualitative to quantitative indicators and from indicators of tacit knowledge to explicit knowledge. Managing and human aspects of TC involve more tacit knowledge; therefore, many of their indicators can be placed in the first and second categories. But hard and information and knowledge aspects of TC have more explicit knowledge; therefore, many of their indicators can be included in the third and fourth categories. After defining the model and its dimensions, through thorough desk research and interviews with automotive industry researchers and experts, indicators of dimensions were determined. We use 28 indicators for assessment of managerial aspects, 25 indicators for assessment of hard aspects, 11 indicators for assessment of human aspects, and 6 indicators for assessment of knowledge and information aspects. Tables 2–5 describe some of dimensions' indicators in detail.

Table 2: Indicators of organizing and managing technological capability.

Indicators	Evaluation Methods
Technology-driven culture in organization (organization perspective on technology)	Qualitative
R&D management and organization	Qualitative
Proportion of R&D costs to sales	Quantitative
Proportion of technical personnel to whole	Quantitative
Proportion of R&D personnel to whole	Quantitative
Level of relationship with academics	Quantitative
Number of patents	Quantitative
Organization flexibility in product development and production	Qualitative
Level of process innovation	Qualitative
Level of product innovation	Qualitative
Level of improvement in product design	Qualitative
Level of organization efforts to implement 5S systems	Qualitative
Level of organization efforts to implementation of quality systems	Qualitative
Level of coordination and integration of production department with other departments	Qualitative
Level of customers' needs identification and compliance	Qualitative
Organization ability in on-time delivery	Quantitative
Strategic planning and its documentation	Qualitative
Procedures for benchmarking and unremitting evaluation of competitors	Qualitative
Implementation of strategy of technology	Qualitative
Technology strategy in relation to organization strategy	Qualitative
Technology acquisition capabilities	Qualitative
Technology transfer capabilities	Qualitative

(*Continued*)

Table 2: (*Continued*)

Indicators	Evaluation Methods
Technology utilization capabilities	Qualitative
Technology protection capabilities	Qualitative
The average time of developing new products (month)	Quantitative
Preventive maintenance culture and systems	Qualitative
Financing and investment in new projects	Qualitative
Contribute margin	Quantitative

Table 3: Indicators of hard aspects of technological capability.

Indicators	Evaluation Methods
Level of equipments complexity	Qualitative
Intensity of labor (labor costs/production costs)	Quantitative
Intensity of capital depreciation (depreciation/production costs)	Quantitative
Intensity of material consuming (material costs/production costs)	Quantitative
Level of overhead costs (overhead costs/production costs)	Quantitative
Proportion of returned products to total production	Quantitative
Indicators of output quality (PPM internal)	Quantitative
Process capability index (CPK)	Quantitative
Reliability of equipments (OEE Indicator)	Quantitative
Set-up time	Quantitative
Level of applying ergonomics in equipments	Qualitative
Level of incidents	Quantitative
Optimized layout (transportation and waiting time/total production cycle time)	Quantitative

(*Continued*)

Table 3: (*Continued*)

Indicators	Evaluation Methods
Type of transportation system (level of automation)	Qualitative
Adequacy of controlling equipments	Qualitative
Using level of process controlling methods and equipments	Quantitative
Novelty level of controlling equipments	Qualitative
Level of defects	Quantitative
Level of energy consuming in planet	Quantitative
Level of diversity use of technology	Qualitative
Level of technology novelty (technology situation in its life cycle)	Qualitative
Level of human errors control (making equipments unerring)	Quantitative
Level of novelty in production process and methods	Qualitative
Level of novelty in products	Qualitative
Completeness of production cycle	Qualitative

Table 4: Indicators of human aspects of technological capability.

Indicators	Evaluation Methods
Personnel education level	Quantitative
Education per capita	Quantitative
Personnel tendency to success	Qualitative
Personnel tendency to teamwork	Qualitative
Personnel familiarity with organization mission and goals	Qualitative
Level of adaptability between personnel job and education	Quantitative
Level of personnel flexibility and change acceptance	Qualitative

(*Continued*)

Table 4: (*Continued*)

Indicators	Evaluation Methods
Personnel job experience on average	Quantitative
Level of operators expertise in applying and maintaining machinery	Qualitative
Level of human resource capabilities	Qualitative
Recommendations per capita (number of suggested, efficient, and implemented)	Quantitative

Table 5: Indicators of knowledge aspects of technological capability.

Indicators	Evaluation Methods
Information updating procedures	Qualitative
Desirable information architecture based on standard methodologies	Qualitative
Level of IT tools utilization (intranet, website, etc.)	Qualitative
PC per capita	Quantitative
Level of information transmission smoothness	Qualitative
Level of information inclusiveness, accuracy, and exactness	Qualitative

Based on these arguments, we proposed our conceptual model and different indicators were selected to assess diverse aspects of TC in automotive parts manufacturers in Iran.

Data Analysis and Results

We implemented the TCA model in 10 automotive parts manufacturers. We assessed the quantitative and qualitative indicators of this model using a questionnaire and gathered data from these companies. Moreover, the qualitative indicators were also assessed by on-site visit and interview with experts and managers of these companies.

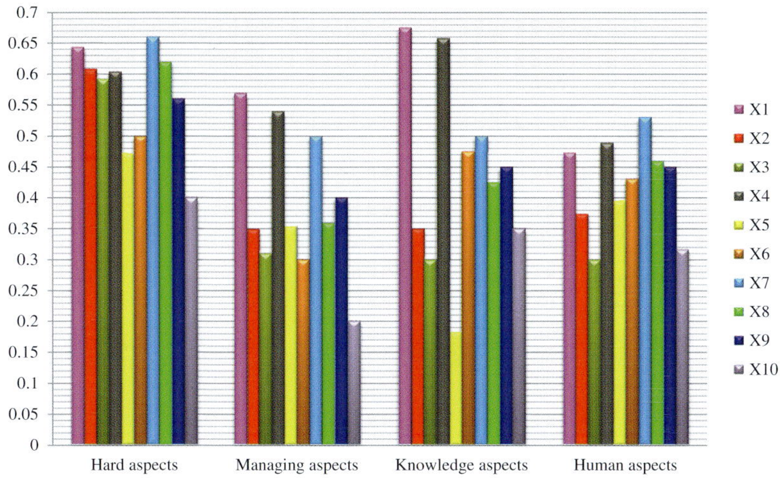

Figure 3: Comparative technological capability of 10 companies in each dimension.

The final results of the implementation phase are shown in Figs. 3 and 4, and Appendix A. Figure 3 shows the comparative TC of 10 companies in each dimension and Fig. 4 illustrates overall comparative TC of these companies. Also, assessment results for each company separately are shown in Appendix A.

As shown in Fig. 3, the first (X1) and the fourth (X4) manufacturers suffer weaknesses regarding human aspects of TC, but are stronger with respect to their capabilities in managing, hard and knowledge aspects. X2, X3, X8, and X9 showed weaknesses in human, managing, and knowledge aspects of TC, but their capabilities in hard aspects are stronger. X5, X6, and X10 have weaknesses in all aspects of TC. Just X7 has smooth TC and strength in all aspects.

Based on the results, the following can be concluded:

- Lack of appropriate soft aspects of TC such as managerial, human, and knowledge aspects, in most cases
- Lack of smooth and balanced TC, in most of cases
- Weaknesses in all aspects of TC, in most of cases
- Lack of appropriate tacit knowledge, in most of cases

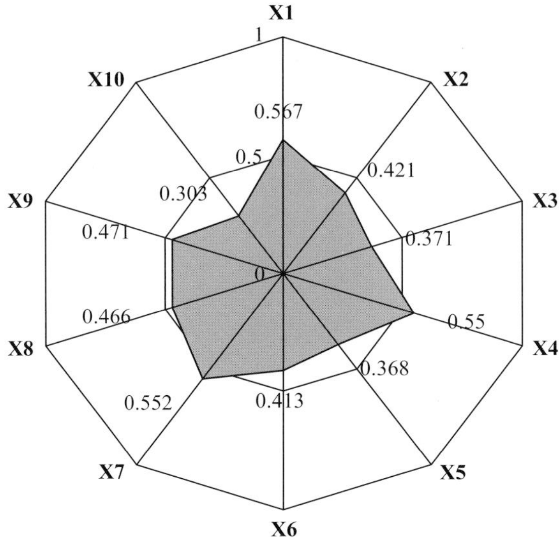

Figure 4: Comparative overall technological capability of 10 companies.

Therefore, this model provides an appropriate framework for assessment of TC in automotive parts manufacturers and graphically demonstrates the most important strengths and weaknesses of these companies.

Conclusions

Having a realistic understanding about technological capabilities in a competitive industry like automotive parts manufacturing industry, is a critical factor in any company's success. In this chapter, diverse approaches to TCA and technology auditing were used to develop a model for assessment of TC. The split of these capabilities in their specific and detailed dimensions was the particular focus of this study.

As a whole, this chapter poses two issues: First, nature and definition of technology and second, different approaches to TCA. Reviewing literature and developing localized and applicable indicators, we proposed a model for TCA in automotive parts manufacturers in Iran.

This model has four dimensions including aspects of management, hardware, human, and information and knowledge. Then, we implemented it in 10 automotive parts manufacturers in Iran. Based on implementation outcomes, we assessed different aspects of technological capabilities in these companies and determined their most important strengths and weaknesses.

Therefore, this TCA model provides an appropriate framework for assessment of TC in automotive parts manufacturers and has several interesting characteristics:

- It takes a holistic view about technology and TCA in an organization and encompasses management aspects, hard aspects, human aspects, and knowledge aspects of TC.
- It encompasses composite indicators of TCA, both qualitative and quantitative.
- It includes both tacit and explicit aspects of technological knowledge.
- It can graphically visualize each TC dimension and the most important strengths and weaknesses.
- It can graphically visualize and determine whether TC dimensions of an organization is smoothed and balanced or not.
- It can assess the evolution of TC dimensions through assessing TC of an organization in different periods.

The main result of this chapter is that TCA models prove to be powerful tools to analyze and compare the technological capabilities of different firms so as to select the most appropriate suppliers in the supply chain. Moreover, after implementing this assessment, automotive manufacturers were assessed, identified their main weaknesses, and selected an appropriate strategy to improve them.

Lack of data was one of the limits of our research. This conceptual model has been developed for automotive parts manufacturing industry. So, future research may pursue discovering a suitable model for assessment of TC in high-tech industries or assessing companies' dynamic capabilities.

Appendix A

Table A.1: Results of the implementation of model in 10 automotive parts manufacturers.

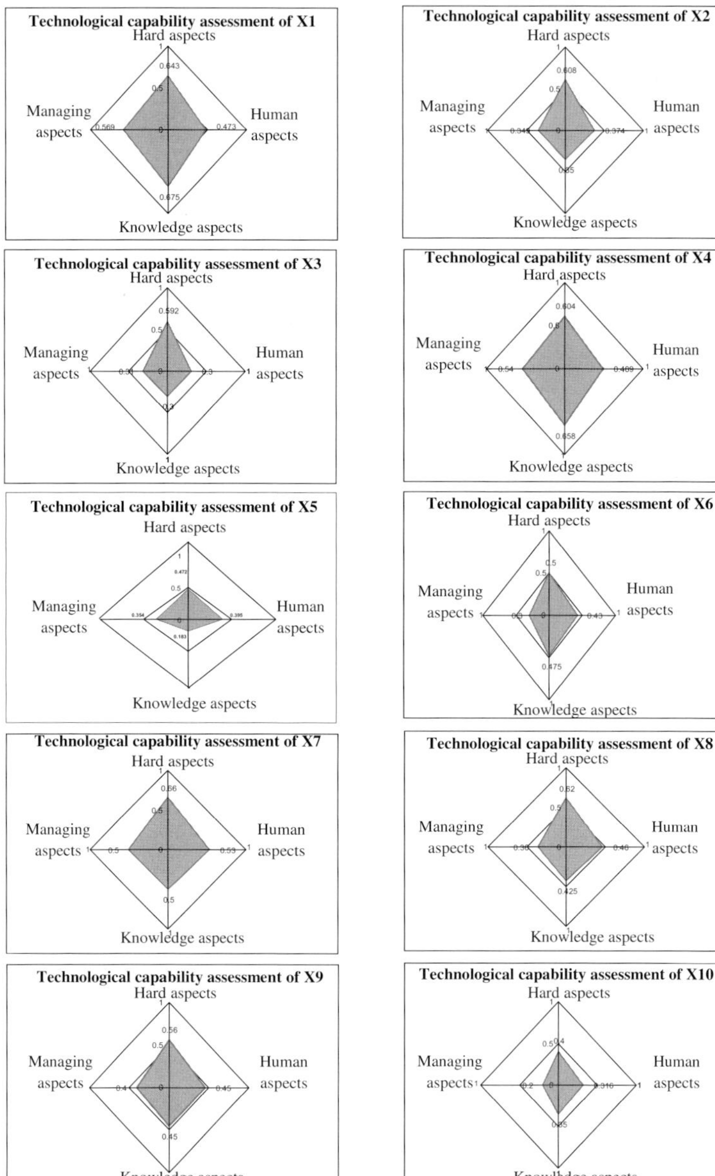

References

APCTT, (1989). *Atlas Technology: A Framework for Technology Planning*. New Delhi: Asian and Pacific Center for Transfer of Technology (APCTT) Publications.

Arasti, M. (2004). A Classification of Methods for Technology Auditing. Proceedings of the PICMET'04, Korea.

Arasti, M. and Pakniat, M. (2006). A Classification of Models for Technology Strategy Formulation. Proceedings of the EUROMOT Conference on Management of Technology, Birmingham, United Kingdom.

Bagheri Moghaddam, N., Mousavi, S. M., Nasiri, M., Moallemi, E. A. and Yousefdehi H. (2011). Wind energy status of Iran: Evaluating Iran's technological capability in manufacturing wind turbines, *Renewable and Sustainable Energy Reviews* **15**: 4200–4211.

Chiesa, V. (2001). *R&D Strategy and Organization*. London: Imperial College Press.

Cowan, M., David, P., and Foray, D. (2001). The explicit economics of knowledge codification and tacitness. *Industrial and Corporate Change* **9**, 211–253.

Coombs, J. E. and Bierly III, P. E. (2001). Looking through the kaleidoscope: Measuring technological capability and performance. *Academy of Management Proceedings*, B1–B7.

Duysters, G. and Hagedoorn, J. (2000). Core competencies and company performance in the worldwide computer industry. *Journal of High Technology Management Research* **11**(1): 75–91.

Ford, D. and Saren, M., 1996. *Technology Strategy for Business*. London: International Thomson Publishing.

Gregory, M. J. (1995). Technology management: A process approach. *Proceeding of the Institution Mechanical Engineers* **209**: 347–356.

Janeš, A. and Dolinšek, S., 2007. Technology audit model (TAM) and the impact of technology on companies and society. Paper presented at the Seventh International Conference on "Enterprise in Transition," Bol-Split, Croatia, 24–26 May.

Johnson, B., Lorenz, E., and Lundvall, B.-A. (2002). Why all this fuss about codified and tacit knowledge. *Industrial and Corporate Change* **11**: 245–262.

Khalil, T. M. (2000). *Management of Technology: The Key to Competitiveness and Wealth Creation*. Boston, MA: Irwin McGraw-Hill.

Kline, S. (2003). What is technology? In *Philosophy of Technology* edited by Robert C. Scharff and V. Dusek. Hoboken, NJ: Blackwell Publishing.

Lall, S. (1990). *Building Industrial Competitiveness in Developing Countries*. Paris: OECD.

Lindsay, J. (1999). *The Technology Management Audit*. UK: Cambridge Strategy Publication.

Lin, M. J. L. (1997). The research on the measurement of corporate technology capability. *International Journal of Technology Management* **13**: 133–139.

Nonaka, I. and Takeuchi, H. (2004). *Hitotsubashi on Knowledge Management*. Singapore: John Wiley & Sons.

Ortega, M. J .R. (2010). Competitive strategies and firm performance: Technological capabilities' moderating roles. *Journal of Business Research* **63**(11): 1273–1281.

Phall, R., Farrukh, C. J. P., and Probert, D. R. (2001). Technology management process assessment: A case study. *International Journal of Operations and Production Management* **21**(8): 1116–1132.

Phaal, R., Farrukh, C. J. P., and Probert, D. R. (2004). A framework for supporting the management of technological knowledge. *International Journal of Technology Management* **27**(1): 1–15.

Panda, H. and Ramanathan, K. (1996). Technological capability assessments of a firm in the electricity sector. *Technovation* **16**(10): 561–588.

Panda, H. and Ramanathan, K., 1997. Technological capability assessment as an input for strategic planning: Case studies at Electricite de France and Electricity Generating Authority of Thailand. *Technovation*, **17**(7): 359–390.

Porter, M. E. (1985). *Competitive Advantage: Creating and Sustaining Superior Performance*. New York: Free Press.

Radosevic, S. (1999). *International Technology Transfer and Catch-up in Economic Development*. Cheltenham: Edward Elgar Publishing.

Štrukelj, P. and Dolinšek, S. (2011). 3-Level Modeling of Organization's Technological Capability. Technology Management in the Energy Smart World (PICMET), Proceedings of the PICMET'11, Portland, OR.

VanWyk, R. (2004). *Technology: A Unifying Code; A Simple and Coherent View of Technology*. Cape Town, South Africa: Stage Media Group.

Zedtwitz, M. and Jin, J. (2004). The Process of Technological Capability Development in the Mobile Telephone Manufacturing Sector: A Case of Eastcom (China), Proceedings of the IAMOT Conference on Management of Technology, Washington, DC.

Zeleny, M. (1986). High technology management. *Human Systems Management* 6, 109–120.

Chapter 6

R&D Portfolio Management: The Intellectual Property Scorecard

*Irfaan Khota** *& Leon Pretorius*†

The model presented in this chapter extends on the findings presented at the Portland International Center for Management of Engineering and Technology (PICMET) (Khota and Pretorius, 2008b). It is also a limited extension of the authors' earlier similar work published in 2012 on the topic (Khota and Pretorius, 2012). The research encompasses *inter alia* (i) strategic considerations affecting producers and manufacturers, (ii) identification of incubators of competitive advantage within the knowledge economy, (iii) the intellectual property (IP) scorecard as a tool to optimize on IP-based value creating organizational activities, and (iv) limited field research.

The research focuses on the inductive development of Kaplan and Norton's balanced scorecard that identifies areas for development in knowledge management, and specifically the required strategic management stance demanded from product developers and innovators with respect to IP, to achieve the competitive high ground. Inductive reasoning in research as used in this research is presented by Cooper and Schindler

*Postgraduate School of Engineering Management, University of Johannesburg, Rizqapital Management, Johannesburg, Gauteng Province, Republic of South Africa.

†Department of Engineering and Technology Management, Graduate School of Technology Management University of Pretoria, Pretoria, Gauteng Province, Republic of South Africa.

(2006) as an effective method especially in the early stages of research and for explorative research.

The sources of economic growth now encompass both the traditional increases in the purchasing power of the population, as well as the explosion and application of new technologies that alter the business models and spending patterns of consumers (Garelli, 2006). In a context where global economic growth is driven by the emergence of new markets, competitiveness depends on the management of multiple competencies in order to exploit resource and knowledge capital, together with the management ability to adapt better and faster than the competition to the dynamic and complex competitive landscape (Garelli, 2006).

The top rankings of Switzerland and the Nordic countries in global competitiveness show that a successful boost in competitiveness in an increasingly complex global economy is attainable through good institutions, competent macroeconomic management, world-class educational attainment, good health care, market efficiency, business sophistication, and a focus on technology, innovation, and investment in research and development (R&D) (WEFORUM, 2006b, 2007). In emerging economies focused on industrial growth as a development strategy, R&D is a requirement for competitiveness (WEFORUM, 2006b), and the basis of competitive advantage and the new core business is the development and deployment of intangible assets (knowledge) (Teece, 2001).

Within the "IP" arena, patents are accepted internationally as a reflection of a country's inventive and technological achievements and are used for monitoring and assessing national systems of innovation (Pouris, 2005).

The Challenge for Innovators and Product Developers Specifically

Globalization, technological advances, changing customer needs, shorter product life cycles, and increased world competition drive new product introductions (Cooper, 2001), where a firm's competitiveness depends on its ability to connect, manage its numerous assets, promote participation in the global network of infrastructure, and leverage the opportunities presented by incremental innovations by entrepreneurs (Garelli, 2006; Hunt, 2013). The management challenge for export growth is thus to expand manufacturing in the current comparative advantage sectors coupled with an expansion of service driven sectors (through investment in transport, technological, and communications infrastructure) (Krakoff, 2003; WEFORUM, 2007).

The Key Strategic Drivers to Consider in Targeting Manufacturing and Innovation Growth

Innovation has been defined as the embodiment, combination, or synthesis of knowledge in original relevant, valued new products, processes, or services (Luecke and Katz, 2003). Competencies in developing innovation, together with efficiency, customer relations, distribution, and technology that aligns with the organization's overall strategy, create value (Ulrich and Smallwood, 2003). The pace of change continues unfettered, and innovation is recognized as a key environmental and contextual change driver (Marton-Lefevre *et al.*, 2003) and a key source for sustainable solutions (UN, 2008).

Good innovation management is a key management lever to business growth and development challenges. Organizational conditions that promote innovation and knowledge creation include: access to capital, flat hierarchies, minimal bureaucracies, and the autonomy to try new things and possibly fail (Teece, 2001; Von Krogh *et al.*, 2000).

From a technology and product development perspective, methodologies should generally adapt with context. Historically, increasing productivity has been driven by strategic management in (i) operational improvement in quality and reengineering, (ii) cost savings through outsourcing, and (iii) taking advantage of national comparative advantages through globalization.

Successful innovation and product development in the new context differ from historical methodologies. Off-shoring and outsourcing are now a key part of the globalization strategy of companies (Barloworld, 2007; Welborn and Kasten, 2005), coupled with a major revolution underway where emerging economies are now competing not only with cheap manpower but also with cheap brainpower driven by the desire, attitude, and ambition to be competitive and thus create wealth (Garelli, 2006; Teece, 2001).

Coming from a leadership paradigm that encourages experimentation in order to learn (Farber, 2004), product development within this context requires effective management strategies that overcome the "innovator's dilemma" by learning to do many things right (Anderson, 2005; Christensen and Raynor, 2003; Christensen, 1997). The impact of the current context of rapid product development on management approach requires inter alia: smaller teams, product simplicity, attention to industrial design, flexible/cost-effective global manufacturing alliances (e.g., China), particular marketing prowess (including physical store, Internet, licensing/distributors, etc.), and legal diligence (Baker, 2008). As Baker (2008) indicates, management need to recognize two significant contextual product development requirements: (i) that each of the five phases of technology/product development (i.e., concept design, design development, detailed development, preproduction, and production) are now considered fundamental, and (ii) that product design, manufacturing, planning, and marketing development need to happen simultaneously rather than sequentially.

Stimulating innovation should be a key management focus area (Marton-Lefevre *et al.*, 2003), and achieving this requires a strategic management approach that is holistic, multidimensional, and targeted. In terms of value considerations that are multifaceted, management must (i) strategize, measure, and manage both tangible and intangible value creation drivers (Ulrich and Smallwood, 2003); (ii) create a positive corporate reputation that influences investor decisions and drives up sales

(Hancock, 2005; Jackson, 2004); (iii) adopt sustainable business strategies that necessarily address value creation from the perspective that it is sensitive to, maps, and manages the evolving context, resources application, and opportunities facing the business environment (Khota and Pretorius, 2008a; Khota *et al.*, 2004); and (iv) appreciate the differences between value creation and value appropriation arising from knowledge-based activities, particularly threats and opportunities presented by entrepreneurs engaged in incremental innovation (Hunt, 2013).

Given the enormous prevailing environmental and contextual challenges that require strategic robust innovation to solve them, design, development, deployment activities across the context, resources, and opportunity arenas must be sensitive to industrial (i.e., technical, market, and financial sensitivities) innovation from the perspective that management's focus must promote inter alia the responsible creation and utilization of new knowledge within the knowledge economy that is manifested responsibly into robust industrial innovations that address challenges from a holistic perspective (a "triple bottom line" management approach (Jackson, 2004; Ulrich and Smallwood, 2003).

Our research suggests nine strategic focus areas spanning across the context-management, resources-management, and opportunity-management arenas (see Fig. 1) that are considered "incubators" of competitive advantage (Khota and Pretorius, 2008a, 2008b, 2009). From a product life cycle perspective, these management focus areas should be encapsulated within an organization's balanced scorecard to ensure strategic priority.

Management needs to learn to make the difficult choices quickly, compromise, and deliberately differentiate, employ tactics such as "first to market", "operational excellence", and "customer relationship management" with a focus on competitiveness through effective implementation of those strategies (through customer centric, process driven, bottom-up, transformational, and leadership-driven management

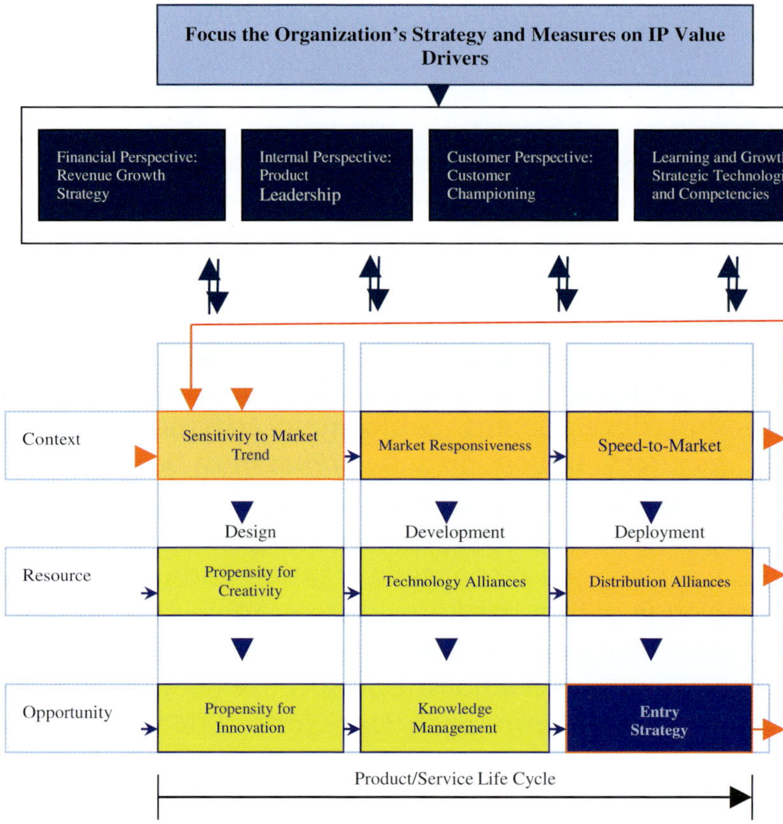

Figure 1: The 3D balanced scorecard paradigm — nine areas of strategic influence (Khota and Pretorius, 2008a, 2008b; Khota *et al.*, 2004).

interventions) (Anderson, 2005; Miyake, 2002). Ultimately, customers are purchasing the utility of a product or service; the most successful companies are those that make it possible for customers to accomplish the functional need more easily, economically, or conveniently (Christensen *et al.*, 2004; Norman, 2005). In this sense, the balanced scorecard concept is particularly useful.

The next section focuses specifically on the importance and impact of good knowledge management and IP as contributors and outcomes of good innovation management.

Context Evolution and Strategic Stance

Three aspects of product design — appearance, performance, and the ideas/memories the item evokes influence consumer's emotional reactions (Norman, 2005). The emerging context and the approaching context over the horizon are to a large degree knowledge driven (Wylie, 2000) (see Table 1), with disruptive innovations (i.e., adequate, smaller, simpler, and cheaper innovations) defining new value propositions of companies in a value network where capabilities are underpinned by resources (people, technology, products, cash, channels, and brands), processes (recruitment, training, R&D, manufacturing,

Table 1: Context trends.

Historic Trend	**Emerging Trend**
Valuing tangibles	Valuing intangibles
Optimization strategies	Innovation strategies
Focus on medium and delivery technology	Focus on content and presentation
Valuing reproducible money and goods	Valuing irreproducible time and infinitely reproducible information
Knowledge filtered by experts	Knowledge filtered by interested parties
Resource restraints	Time restraints
High entry costs	Low entry costs
Equilibrium and stability	Sustainability and adaptation
Current Trend	**Over the Horizon**
Virtual "reality"	Post virtual realities
Focus on elements	Focus on context
Knowledge generation and personal expertise	Knowledge utilization and group understanding
Focus on text centered literacy	Focus on multimedia and aural communication

(*Continued*)

Table 1: (*Continued*)

Solving problems	Seeking opportunities
European and African lack of innovation	European and African innovation leadership
Economies of scale	Economies of scope
Additive measurement	Combinatory measurement
Deregulation and freedom	Reregulation according to agreed principles

Source: Adapted from Christensen and Raynor (2003) and Wylie (2000).

market research, and budgeting), and values (ethics, customer demands, cost structures, risk tolerance, and emphasis on new opportunities) (Christensen and Raynor, 2003; Christensen *et al.*, 2004; Teece, 2001). Paradoxically, great companies do not fail because they do bad things, but rather they fail because they keep doing better at the things that made them great (Christensen, 1997). The successful companies are those that are able to recreate themselves to meet new competitive demands (Welborn and Kasten, 2005), where success is driven by: (i) the adaptation of resources, processes, and values with the contextual trends introducing disruptive innovations (Christensen *et al.*, 2004; Teece, 2001), (ii) where the best designs are driven by dynamic, flexible processes (Norman, 2005), (iii) a competency to introduce products quickly, which creates competitive advantages and higher profitability (Cooper, 2001), and (iv) resource allocation that prioritizes projects based on cost–benefits analysis, maximization of R&D, and new product exploration (Cooper, 2001). These are viewed increasingly from the perspective of value-chain impact and virtual communities, where success also depends on collaboration/contribution, resource focus, achieving critical mass, and matching community and business needs (Spaulding, 2010).

Intangible assets (staff skills, strategic and process quality, patents, copyright and software, trademarks/brands, supplier and customer relationships, etc.) are currently the key factors of production driving competitiveness (Hofmann, 2005).

Since much of the market valuation of companies today, particularly technology companies, is based on intangible assets (brands, IP, people) compared to the historical focus on tangible assets (property, plant, and equipment), a management focus on leveraging and capitalizing on these assets is justifiable (Fairchild, 2002; Hofmann, 2005; Howrey, 2001). In addition, globalization introduces increased vulnerability and necessity to protect proprietary information from being copied or stolen (Garelli, 2006).

In five separate field surveys conducted by the authors presented in earlier papers (Khota and Pretorius, 2009; Khota, 2008a) to assess what the product development and business evaluation criteria are that successfully attract the interest of investors, it was found that (i) IP is indeed an increasingly important consideration by investors and lenders, and (ii) product development companies that are sensitized to these specific IP considerations are considered to be better placed to compete, generate profits, and attract funding from external investors. More specifically, it was found that:

- Most investors consider patents, trademarks, and copyrights to be important IP
- A product's IP profile is an important consideration for outside Investors
- A product's IP strategy influences the investment decisions
- A greater emphasis will be placed on IP considerations by investors into the future
- Many companies currently have inadequate IP strategies in place

An earlier survey conducted by Howrey (2001) among corporate analysts, yields that an effective IP strategy should encompass:

- Patent protection (which justifies R&D spend)
- The use of IP to protect market position and prevent competitors from entering the market

- An approach which balances the emphasis on IP and tangibles
- Exploitation of potential
- Generating royalties through the patent portfolio

The following are cited as economic reasons for an organizational management strategy that focuses on knowledge (Knapp, 1998):

- The global perception of knowledge as a key asset
- The rise of occupations that create and use knowledge
- The convergence of information and communication technologies

Apart from technology, other organizational knowledge "enablers" include content, learning, culture, and leadership (Knapp, 1998). Knowledge is increasingly accepted as a source of value creation and a differentiator between firms and is based on knowledge-use rather than knowledge-possession (Breunig and Roberts, 2014).

From a strategic planning perspective, having a clear and effective IP strategy is thus deemed fundamental to a company's success and survival in the marketplace, and those who lack an effective IP strategy are generally not expected to become market leaders (Howrey, 2001). Having a strategic tool, such as the IP scorecard presented here, is considered an essential management aid in focusing attention on development and commercialization of technology innovations from an IP-centric perspective.

Organizational Empowerment through Knowledge Management

Knowledge in organizations is seen in complex and broad terms including structured data, patents, programs and procedures, intangible knowledge and capabilities of people, and novel and unique ways of doing business (Fairchild, 2002), where knowledge management encompasses the processes of innovation, knowledge creation, knowledge acquisition, organization, application, sharing, and replenishment, all geared toward creating value from intellectual capital (Knapp, 1998).

The knowledge-centric organization thus seeks to create value by leveraging the following capital resources within the organization, which jointly provide a basis for knowledge creation and utilization (Fairchild, 2002):

- Human capital: all individual capital (knowledge, skills, experience)
- Intellectual capital (includes intangibles within knowledge, information, and skills that can produce an asset of equal or greater importance than land, labor, and capital)
- Structural capital (includes processes, structures, and systems)
- Social capital (goodwill between people, which promotes sharing, arising from physical and virtual interchanges)

Firms can stimulate its members to participate in knowledge creation and the sharing of products or services by leveraging and enhancing social capital (i.e., relational networks) (Sheng and Hartono, 2015); effective knowledge management, which justifies further management expenditure and attention, encompasses measurable benefits of knowledge implementation and the ability to exploit existing technology (Knapp, 1998). As an example, IBM is a good case of effective knowledge management: The US Patent and Trademark Office's annual league table of corporate patent filing reveals that in 2000, IBM filed 2886 patent applications — the income from licensing IP is one of the IBM's fastest-growing sources of profit ($1.7bn: FY 2000) (IBM, 2004).

Fairchild (2002) cites four management perspectives emanating from knowledge: the value of knowledge from a balance sheet perspective, the market and technological exploitation of IP, the capturing of project-based learning, and the management of knowledge workers (Knapp, 1998). Efficient knowledge utilization may be pursued by expressing knowledge in ways that promote monetization; value-creation through knowledge resources within organizations requires that knowledge is deployed and utilized via purposeful knowledge communication (Breunig and Roberts, 2014). The process of creating and sharing knowledge drives value creation by effectively accelerating the development of new products, enhances relationships with partners, increases contextual involvement, increases customer engagement, and it intensifies product and service innovation (SheSng and Hartono, 2015).

While protecting and exploiting IP is the key to success, investors recognize that companies with significant IP assets have a competitive advantage, are likely to be more profitable and are more likely to survive a downturn (Howrey, 2001).

There is thus definite potential with respect to IP development and value creation, through appropriate IP strategy and management.

As further validation, the research conducted by the authors among venture capital investors (Khota, 2008b) indicates that there is a poor inclination toward technology/product development and commercialization investment opportunities as reflected in Table 2.

The next section discusses the IP scorecard, a strategic management tool emanating through inductive reasoning from the exploratory research. This tool focuses management attention on strategic IP development and management with a view to value creation through IP.

The IP/Knowledge Scorecard: A Conceptual Framework

As a strategic management tool, the balanced scorecard method is used by many companies to define the organization's strategy, create targets and initiatives for implementing the strategy, measuring the results of such implementation, and creating feedback mechanisms for improving the strategy (IBM, 2004;

Table 2: Survey conducted among seven venture capital investors: Summarized results.

There is a little or no novelty in the product/technology offering (i.e., lack of IP). Investors generally prefer supporting a product that displays clear novelty which enables competitive advantage based on its innovative attributes.
The opportunity fails to boast significant technology competitive advantages (i.e., lack of IP). In general, the greater the product's technology advantage the lower the perceived market risk by investors.
The opportunity fails to exhibit adequate IP protection claims whether formally or via trade secrets for the technology/products being developed (i.e., a poor product protection matrix structure surrounding the product). Investors also consider it essential in high-technology investments that the inventor or technology developer plays an active and committed role in the new business venture. The findings of the IP survey (e.g., that investors consider the product's IP and the robustness of the IP strategy) and the literature study findings also suggest that the ranking of IP as an important criterion should escalate with time.

Kaplan and Norton, 1996). From an innovation management perspective, while there are numerous management metrics within the financial and customers' areas of the balanced scorecard (Fairchild, 2002; Kaplan and Norton, 1996), the current research seeks to explore metrics specifically within the internal and learning and growth areas that create shareholder value.

Industry experts accept that learning and growing organizations are better positioned for strategic success into the future — the learning and growing organization is one in which knowledge management activities are deployed and expanding in order to leverage the creativity of all the people in the organization (Kaplan and Norton, 1996; Muralidhar, 2005).

In general, investors agree that better corporate information and strategic guidance on the company's intangible assets and their development or commercialization should directly impact on corporate valuations and the financing opportunities available as a result thereof (Hofmann, 2005; Howrey, 2001; Khota, 2008a; Khota and Pretorius, 2009; Lev, 2004).

Despite the limitations of accounting regulations, sensitivities toward protecting "company secrets" and the challenges surrounding exploiting intangible asset value (e.g., comparability, increased risk, valuation methods, and market skepticism), there is a real benefit for companies who do strategic planning around intangible assets (Hofmann, 2005). Managing and creating "balance sheet" value from intangible assets should lead to more growth through: capital allocation optimization, investor confidence as a result of better information on intangible assets, and the appropriate market valuation of the company (Hofmann, 2005). Enhancing the recognition of intangible assets as a viable form of collateral, through improved IP management and structuring, licensing, and so on, will improve the value attributed to such assets (Hofmann, 2005; Teece, 2001).

Ultimately the monetary valuation of intangible assets will be guided by the following considerations: financial objective (loan, investment, or disposal), granularity (appraisal of individual intangibles [e.g., specific patents], knowledge intensive projects, whole business divisions, or entire companies), and valuation perspective (internal or external valuator) (Hofmann, 2005). While the cost price, market value, or income approaches may be used to value an intangible asset, the income approach is preferred most, since this method takes cognizance of the forecast earnings attributable to intangible assets specifically and uses the discounted cash flow (DCF) analysis on these future intangible driven earnings to indicate present value (Hofmann, 2005). Lev (2004) provides more guidance on the actual valuation methodology of intangible assets.

In terms of strategic focus areas for value optimization of IP, two key activities are necessary (Hofmann, 2005):

- Intangible assets that are of particular relevance to the company-specific value-added processes must be identified and analyzed

- The transmission methods by which central intangible assets are transformed into future revenues must be well understood

Ultimately, however, an effective IP strategy is necessary. Conversely, an ineffective IP strategy consequentially results in an opportunity cost and R&D inefficiencies that investors and external stakeholders frown upon or competitors take advantage of to the detriment of the organization — Howrey (2001) cites British Telecom (BT) as such an example, losing massive opportunity by owning 15,000 patents protecting more than 1700 inventions and yet failing to capitalize on these effectively.

The Balanced Scorecard

The balanced scorecard is a strategic management tool that promotes management attention metrics, data collection, and analysis from four key perspectives (i.e., learning and growth perspective, business process perspective, customer perspective, and financial perspective), aimed at motivating the organization and creating alignment between strategy and outcomes (Arveson, 1998; Kaplan and Norton, 2001). In general, organizations that make their strategy the central organizational agenda with incredible strategic focus and a mobilization of staff to act in fundamentally different ways, guided by the strategy, display exceptional performance (Kaplan and Norton, 2001; Miyake, 2002; Welborn and Kasten, 2005). Key management activities of the balanced scorecard users who displayed exceptional performance have focused on implementing the strategic map (leadership, enterprise-wide operational strategic focus, organizational strategic alignment, personnel strategic commitment, and strategic evolution) (Kaplan and Norton, 2001; Miyake, 2002), and aligning the organizational map with the scorecard of customers and alliance partners (Miyake, 2002).

The balanced scorecard can be the foundation for the strategic management of information and knowledge in organizations (Martinsons *et al.*, 1999), under the management category of learning and growth (Fairchild, 2002). Product developers may refine the balanced scorecard even further by applying a "Product/Innovation Scorecard" (Khota and Pretorius, 2008b) that defines the strategy and performance measures for products and product innovations, and which provides strategic focus and execution for successful product evaluation, adoption, positioning, life-cycle management and growth management (Sahu, 2004b). The strategy-focused product innovation process encompasses: (i) mobilizing product innovation through leadership; (ii) translating the product innovation strategy into product development; (iii) aligning product design and development processes to the product innovation strategy; (iv) making product innovation strategy every product engineers job; and (v) making product innovation strategy a continuous process (Kaplan and Norton 2001; Sahu, 2004b) (see Fig. 2).

The conceptual IP scorecard presented is the result of the literature review and inductive reasoning employed within the nine focal areas identified in the 3D balanced scorecard (Khota and Pretorius, 2008a; Khota and Pretorius, 2008b; Khota *et al.*, 2004) mentioned earlier and the IP product protection matrix (Khota and Stern, 2005).

The Key Strategic Management Focus Areas Driven by IP and Knowledge Management

Within the knowledge economy, knowledge "ownership" and capitalization are measures of success for management. This economy thus warrants the prioritization of the "Internal Drivers" (indicated in Fig. 3) that improve and promote knowledge and IP ownership and utilization, and customer-centric knowledge-sharing aspects (i.e., the knowledge-based aspects

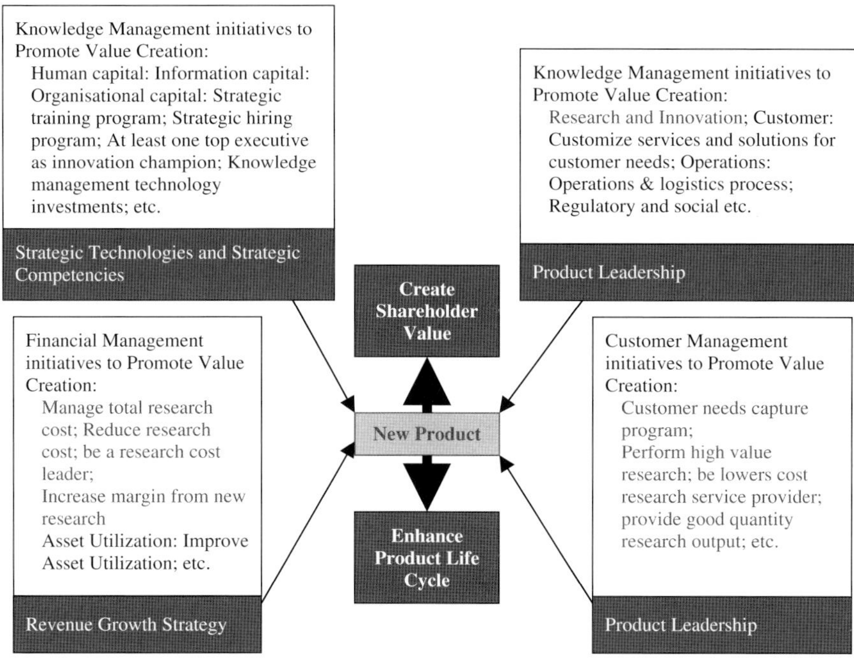

Figure 2: BSC/R&D strategy map — focusing strategy on IP and product innovation (Adapted from Coburn, 2006; Fairchild, 2002; Govindarajan and Trimble, 2005; Kaplan and Norton, 2004; Johannson, 1996; Mouritsen *et al.*, 2001; Sahu, 2004a, 2004b; Spaulding, 2010; Stalk and Lachenauer, 2004; Teece, 2001; Von Krogh *et al.*, 2000; Khota and Pretorius, 2012).

of the "Customer Drivers" indicated in Fig. 3), and the direct impact of these metrics on the other perspectives. Using these prioritized metrics in conjunction with an "IP Product Protection Matrix" approach (Khota and Stern, 2005) that seeks to economically exploit all the IP characteristics of a product to enhance its value (including patent, copyright, design, and trademark), this value-driven portfolio management approach to IP and product development should ensure strategic leadership within the knowledge economy.

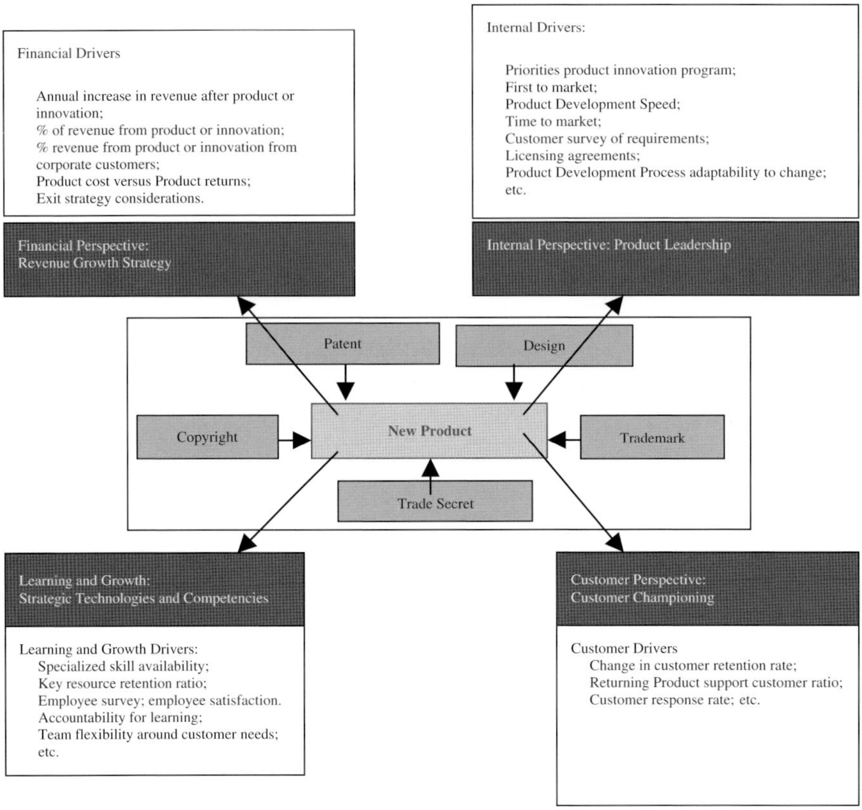

Figure 3: The knowledge-centered BSC/product development metrics (Adapted from Fairchild, 2002; Hindle and O'Connor, 2004; Khota and Stern, 2005; Mouritsen *et al.*, 2001; Sahu, 2004b).

In general, the rate of failure for high-tech products is high, due to a poor rate of acceptance of new technology by consumers (Coburn, 2006). As a management innovation measure, instead of focusing only on new features and low costs, companies developing high-tech products should focus on what their customers think and feel about their products and tailor product innovation to address the needs of this market that is, the "change function", which says that the customers' desire for the product must outweigh their reluctance to adapt to something new", must drive innovation, rather than "Moore's law" or "Grove's law" (Coburn, 2006).

Revolutionizing a business requires strategic innovations (which encompasses unproven business models) through strategic experiments, which are hard to predict and budget for (Govindarajan and Trimble, 2005), with investments in R&D that necessarily seek to promote continual learning, growth and development of internal processes in order to satisfy customer needs and to achieve the financial goals of shareholders (Sahu, 2004a). Strategy maps for the R&D organization include the management activities within the balanced scorecard indicated in Fig. 2.

Businesses must use knowledge management to survive and advance (Von Krogh *et al.*, 2000). The value of intangible assets is achieved with how these assets are aligned with the strategy of the R&D initiatives (Sahu, 2004a), which encompass consideration of strategic job families (i.e., jobs with greatest impact on the strategy), strategic IT portfolio (i.e., high-impact technology investments), and organizational change agenda (i.e., organizational cultural change mechanisms), targeting either complete research solutions, low total cost R&D, and/or research leadership (Fig. 3 provides a holistic view of R&D BSC considerations).

A key consideration from an investor perspective is the optimum timing to license, abandon, or exit a knowledge-based investment/consideration to realize value and reduce risk. Registered IP can be expensive to maintain, particularly global assets where annual renewal fees are in hard currency; a consideration that particularly impacts on enterprise activities within the emerging economies. Research suggests that medium-term economic growth and ongoing innovation are generally characterized by economies with serial entrepreneurial cultures that frame business and product commercialization opportunities with a defined "exit" (i.e., trade sale) in mind

(Hindle and O'Connor, 2004); an exit strategy should form part of the IP scorecard considerations.

In general, the implications emanating from the field research and literature study suggests several key areas where management activity should be focused. These include *inter alia* that (i) management should create incentives for new product development that creates value through IP generation; (ii) that management should encourage third-party collaboration that seeks to generate IP rights and that are protected; (iii) that management should be sensitized to the intrinsic value of its organizations dormant IP and its commercial potential; (iv) that management should create a "freedom to operate" paradigm: freedom to sell, develop, grow in any country, and any place; (v) that management should strategically target creating and exploiting IP in its primary markets; and (vi) that management should manage the pace of change through continual training, new skills acquisition, capex/R&D, and targeting/retaining people of value.

With respect to strategic management of IP, management should seek to (i) develop new IP and protect existing IP; (ii) pursue patenting/and formal registration opportunity arising from innovative designs; (iii) develop an "effective" IP development strategy; (iv) have a very strict attitude on IP infringement; and (v) avail all resources (skills, capital, instruments, monitoring) to develop and protect IP in broad terms.

Conclusion

The knowledge economy demands a strategic management focus on knowledge-based outcomes. The research supports a holistic IP-centric value creation approach to strategic product development that effectively manages all stakeholders' interests. A focus on innovation as a growth and development strategy is imperative for long-term leadership and value creation.

The frameworks and scorecards reflected in this chapter have been successfully and conceptually developed through exploratory research, inductive reasoning, and limited field research reported on and published previously.

Through a management focus on the nine management areas broadly defined as the 3D-balanced scorecard paradigm within the product development life cycle, the context, resource, and opportunity arenas are expected to be optimized toward the strategic intent and leadership. Delivering on that management focus requires the harvesting of knowledge and IP as the key ingredient in the knowledge economy. This necessarily mandates in the first instance a product protection matrix approach to product development. This also necessitates a management convergence and obsession of resources on identifying, codifying, collaborating, and capitalizing on IP and knowledge creation. This should ultimately be recognized through a significant increase in the number of "intangible asset" line entries that appear on the balance sheet, representing the value created through good IP management. Converging on such strategic management through tools such as the IP scorecard (indicated in Fig. 2) must seek ultimately to deliver on the key management question: "Are we creating and leveraging IP to create and enhance stakeholder value"? The IP scorecard also sensitizes management on where and how to carve out competitive turf, particularly from a knowledge and IP value creation perspective, seeking ultimately to enhance corporate/shareholder and product life cycle value.

This research has indicated that IP as a value creation mechanism needs to be a top management priority in a context where investors increasingly look toward good IP management as an indicator of future organizational growth, and where competitive positioning thrives on innovation. The global financial turmoil, which has deepened recently and the increasing focus on better utilization of organizational resources render. The IP scorecard is more relevant as a strategic management tool since

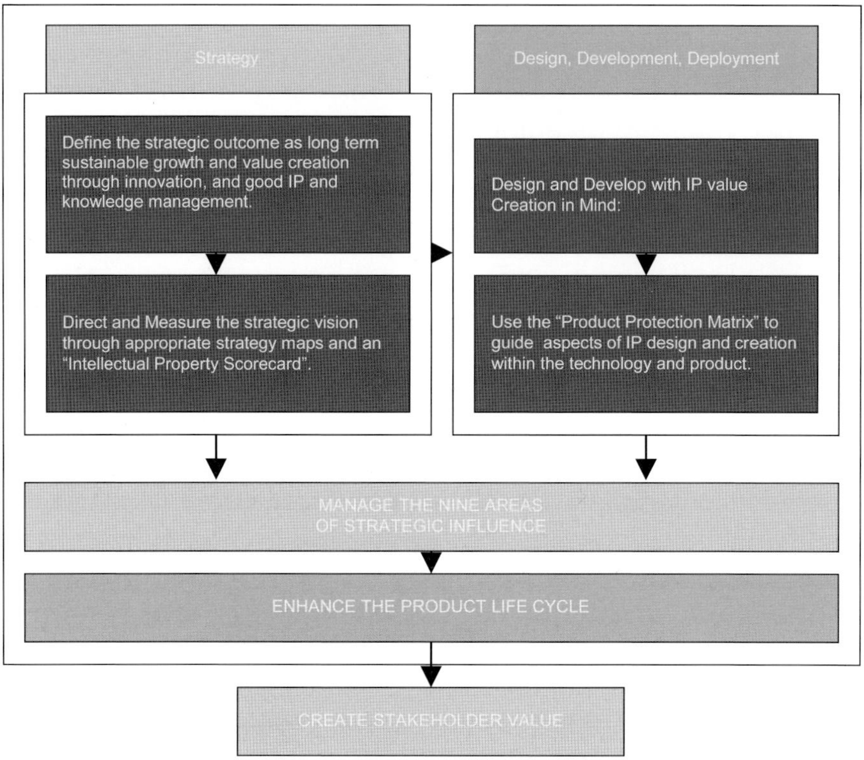

Figure 4: Strategic framework for harvesting IP-based value potential.

it encourages and enables management to drive a robust industrial growth strategy that is explicitly positioned within the knowledge economy, identifying target areas for organizational development around knowledge and IP management, in order to achieve the competitive higher ground, by strategically creating and capitalizing on value arising from innovation and R&D activities.

IP capitalization and investment in R&D have declined in recent years, particularly with respect to emerging nations within the global IP environment, which limits accelerated industrial growth. In most emerging markets, business ventures and their management are challenged to address multiple competencies better and simultaneously, so that they are able to optimize the utilization of organizational resources, knowledge, and management ability, so that they may adapt faster than the competition. Our research has suggested a formalized strategic approach to IP, knowledge and innovation activities.

The management strategic priority is that whatever the IP activity being undertaken (including licensing out of knowledge assets, technology transfer agreements, cross-licensing, etc.), ultimately companies must seek to recover and enhance value from knowledge, IP, and innovation, across the product life cycle in order to carve out financial success. The timeous strategic management of all knowledge-based assets is critical; disruptive and incremental innovation from the market continually threatens such value-potential and tactics to monetize and/or exit such assets timeously should form part of the priority resource allocation considerations by executive management of firms. This requires a coherent strategic management focus on IP, encompassing *inter alia* utilizing IP-centric strategic management tools (including the IP scorecard) to focus management activities on value creation across and within the context, resources, and opportunity arenas (see Fig. 4).

References

Anderson, D. (2005). *If You Don't Make Waves You'll Drown — 10 Hard-Changing Strategies for Leading in Politically Correct Times*. Hoboken, NJ: John Wiley & Sons.

Arveson, P. (1998). What Is the Balanced Scorecard. http://www.balancedscorecard.org/basics/bsc1.html. Accessed 09 November 2006.

Baker, P. (2008). *From Concept to Consumer — How to Turn Ideas Into Money*. Upper Saddle River, NJ: FT Press.

Barloworld. (2007). *Supply Chain Foresight Survey–2007*. Chennai, India: Barloworld Logistics. http://www.barloworld-logistics.com.

Breunig, K. J. and Roberts, H. (2014). Putting your money where your mouth is: Monetising knowledge using communication roles. *European Conference on Knowledge Management: ECKM2014* (Vol. 3, p. 156).

Christensen, C. M. (1997). *The Innovator's Dilemma — When New Technologies Cause Great Firms to Fail*. Boston, MA: Harvard Business School Press.

Christensen, C. M. and Raynor, M. E. (2003). *The Innovator's Solution — Creating and Sustaining Successful Growth*. Boston, MA: Harvard Business School Press.

Christensen, C. M., Anthony, S. D. and Roth, E. A. (2004). *Seeing What's Next — Using the Theories of Innovation to Predict Industry Change*. Boston, MA: Harvard Business School Press.

Coburn, P. (2006). *The Change Function — Why Some Technologies Take Off and Others Crash and Burn*. New York, NY: Portfolio Division, Penguin Group.

Cooper, D. R. and Schindler, P. S. (2006). *Business Research Methods*, 9th Edition. New York, NY: McGraw-Hill.

Cooper, R. G. (2001). *Winning at New Products — Accelerating the Process from Idea to Launch*, 3rd Edition. New York, NY: Perseus.

Fairchild, A. M. (2002). Knowledge Management Metrics via a Balanced Scorecard Methodology. *Proceedings of the 35th Hawaii International Conference on System Sciences*. http://csdl2.computer.org/comp/proceedings/hicss/2002/1435/08/14350243.pdf. Accessed 13 November 2006.

Farber, S. (2004). *The Radical Leap — A Personal Lesson in Extreme Leadership*. Chicago, IL: Dearborn.

Garelli, S. (2006). The World Competitiveness Landscape in 2006. IMD World Competitiveness Yearbook 2006. IMD World Competitiveness Centre. http://www.imd.ch/research/centers/wcc/index.cfm. Accessed 09 November 2006.

Govindarajan, V. and Trimble, C. (2005). *Ten Rules for Strategic Innovators — From Idea to Execution*. Boston, MA: Harvard Business School Press.

Hancock, J. (2005). *Investing in Corporate Social Responsibility — A Guide to Best Practice, Business Planning & the UK's Leading Companies*. London: Kogan Page Business Books.

Hindle, K. and O'Connor, A. (2004). *WESTPAC GEM Australia — A Study of Australian Entrepreneurship in 2004*. Global Entrepreneur

Monitor 2004; Australian Graduate School of Entrepreneurship. Australia: Swinburne University of Technology.

Hofmann, J. (2005). *Value Intangibles! — Intangible Capital Can and Must Be Valued — Owners and Valuers Alike Will Benefit*. Germany: Deutsche Bank Research. http://www.dbresearch.com. Accessed 19 October 2005.

Howrey. (2001). A Survey of Investor Attitudes on IP Protection. http://www.howrey.com/docs/UK_IP_Survey0102.pdf. Accessed 08 January 2007.

Hunt, R. A. (2013). Entrepreneurial tweaking — An empirical study of technology diffusion through secondary inventions and design modifications by start-ups. *European Journal of Innovation Management*, **16**(2): 148–170.

IBM. (2004, September). *Using a Balanced Scorecard to Tie the Result Act to Your Day-to-Day Operational Priorities*. IBM Business Consulting Services, Institute for the Study of Public Policy Implementation. http://spa.american.edu/isppi/documents/leadership_pres1_2004-09-28.pdf. Accessed 21 November 2006.

Johannson. (1996). Human Resource Costing and Accounting. http://www.sveiby.com.au/IntangAss/OECDartUlfjoh.htm. Cited in (Fairchild, 2002).

Kaplan, R. and Norton, D. (1996). *Translating Strategy into Action*. Boston, MA: Harvard Business School.

Kaplan, R. and Norton, D. (2001). *The Strategy-Focused Organisation*. Boston, MA: Harvard Business School Press.

Kaplan, R. S. and Norton, D. P. (2004). *Strategy Maps — Converting Intangible Assets into Tangible Outcomes*. Boston, MA: Harvard Business School Press.

Khota, I. (2008a). *IP SURVEY: A Survey of South African Investor Attitudes on IP Protection*. South Africa: University of Johannesburg.

Khota, I. (2008b). *Survey/Interviews Conducted amongst Venture Capital Investors. Venture Capital Strategic Business Unit*. South Africa: Industrial Development Corporation of South Africa.

Khota, I. and Pretorius, L. (2008a). Embedded Innovation — Strategic Management Incubators for Knowledge Hegemony. *South African Journal of Industrial Engineering*, **19**(1): 105–118. http://findarticles.com/p/articles/mi_qa5491/is_200805/ai_n25501283.

Khota, I. and Pretorius, L. (2008b). Intellectual property scorecard: Strategically capitalising on value created by innovation and R&D.

Proc. PICMET. *Portland International Conference*, **27–31**: 584–592. July. http://www.picmet.org/new/conferences/2008/bulletin08.pdf. Accessed 25 June 2009.

Khota, I. and Pretorius, L. (2009). Sustainable Change-Driven-Innovation: A Private Equity/Investor and IP Hierarchy Perspective for New Product Development Companies. *South African Institute of Industrial Engineering Conference*, October, South Africa.

Khota, I. and Stern, A. (2005). Leveraging intellectual property for strategic product development advantage. *South African Journal of Information Management*, **7**: 4. December.

Khota, I. *et al.* (2004). Strategic Positioning for New-Context Product and Service Leadership. *South African Institute of Industrial Engineering Conference*, September. Indaba Hotel. http://www.saiie2009.co.za. Accessed 11 January 2010.

Khota, I. and Pretorius, L. (2012). Harvesting IP-based value potential: The intellectual property scorecard as a mechanism to capitalize on technological innovation, knowledge, and IP. *International Journal of Innovation and Technology Management*, **9**(2): 1250009. doi:10.1142/S0219877012500095

Knapp, E. (1998). Knowledge management. *Business & Economic Review*, **44**, 4. July–September.

Krakoff, C. (2003). Key Potential Export Markets and the Market Access Barriers Facing Southern African Exporters. The Services Group, South African Global Competitiveness Hub. http://www.satradehub.org. Accessed 08 November 2006.

Lev, B. (2004). Sharpening the intangibles edge. *Harvard Business Review*, **82**(6) June: 108–116.

Luecke, R. and Katz, R. R. (2003). *Managing Creativity and Innovation*. Boston, MA: Harvard Business School Press.

Martinsons, M *et al.* (1999). The balanced scorecard: A Foundation for the strategic management of information systems. *Decision Support Systems*, **25**: 71–88.

Marton-Lefevre *et al.* (2003). Invention and Innovation for Sustainable Development. Report of Lemelson-MIT Program and LEAD International, School of Engineering, Massachusetts Institute of Technology. http://web.mit.edu/Invent/n-pressreleases/downloads/sustainable.pdf. Accessed 09 June 2009.

Miyake, D. (2002). Implementing Strategy with the Balanced Scorecard: An Introduction to the Strategy-Focused Organization. http://www.dmreview.com. Accessed 11 November 2006.

Mouritsen, J. *et al.* (2001). Reading an Intellectual Capital Statement: Describing and Prescribing Knowledge Management Strategies. *Proceedings of the 4th Intangibles Conference*, May. Stern School of Business, New York University.

Muralidhar, S. (2005). The Balanced Scorecard and Knowledge Management. http://www.deccanherald.com/deccanherald/Jul132005. Accessed 13 November 2006.

Norman, D. (2005). *Emotional Design–Why We Love (or hate) Everyday Things*. New York, NY: Basic Books.

Pouris, A. (2005). Technological performance judged by America patents awarded to South African inventors. *South African Journal of Science*, **101**: 221–224.

Sahu, P. (2004a). Describing Research and Development Using Strategy Maps. *IEEE-UT Engineering Management Conference*. University of Texas.

Sahu, P. (2004b). Strategy-focused product innovation with product scorecard and product innovation scorecard. *IEEE-UT Engineering Management Conference*. University of Texas.

Sheng, M. and Hartono, R. (2015). An exploratory study of knowledge creation and sharing in online community: a social capital perspective. *Total Quality Management*, **26**(1): 93–107. Department of Business Administration, National Taiwan University of Science and Technology.

Spaulding, T. J. (2010). How can virtual communities create value for business. *Electronic Commerce Research and Applications*, **9**(1): 38–49.

Stalk, G. and Lachenauer, R. (2004). *Hardball–Are You Playing to Play or Playing to Win?* Boston: Harvard Business School Press.

Teece, D. J. (2001). *Managing Intellectual Capital — Organisational, Strategic and Policy Dimensions*. Oxford, UK: Oxford University Press.

UN. (2008). *Innovation for Sustainable Development — Local Case Studies from Africa*. New York, NY: United Nations. http://www.un.org/esa/sustdev/publications/africa_casestudies/index.htm. Accessed 16 June 2009.

Ulrich, D. and Smallwood, N. (2003). *Why the Bottom Line Isn't! — How to Build Value through People and Organizations*. New York, NY: John Wiley & Sons.

Von Krogh, G. *et al.* (2000). *Enabling Knowledge Creation — How to Unlock the Mystery of Tacit Knowledge and Release the Power of Innovation*. Oxford, UK: Oxford University Press.

WEFORUM. (2007). Global Competitiveness Report 2006–2007. World Economic Forum. http://www.weforum.org. Accessed 08 November 2006.

WEFORUM. (2006b). Technologies for African Growth. World Economic Forum on Africa 2006. 1 June. http://www.weforum.org/en/knowledge/Events/2006/WorldEconomicForumonAfrica. Accessed 08 November 2006.

Welborn, R. and Kasten, V. (2005). *Get It Done! — A Blueprint for Business Execution*. New York, NY: John Wiley & Sons.

Wylie, J. (2000). The Economics of Intangible Value — Trends. Trend Monitor International. http://www.entovation.com/assessment/trends.htm. Accessed 01 July 2008.

CHAPTER 7

R&D Portfolio Management: Real Options Theory in Defense Projects

Won-Joon Jang & Jeong-Dong Lee†*

The capability of developing defense critical technology and its own weapon system is the most important factor for a country's own national security reasons. The United States, which is a superpower in the world today, spends more than 465 billion dollars as an annual defense budget in 2006 to maintain its dominant military power in the world (Ministry of National Defense [MND], 2006). Other defense advanced countries, especially Japan and Israel, also spend billions of dollars in rebuilding their military strength for their own security reasons, too.[1] Moreover, the US Department of Defense (DoD) allocated over 40 billion dollars as a defense R&D (Research and Development) budget, which is over 13% of the total defense budget, for the development of cutting edge defense technology and high-end weapon systems (MND, 2006). Due to the reluctance of transferring defense critical technologies to even friendly allies and its huge costs from acquisition to maintenance, it is obvious that it

*Defense Agency for Technology and Quality, Seoul, Republic of Korea.
†TEMEP, Seoul National University, Seoul, Republic of Korea.

[1] In 2006, Japan spends over 45.1 billion dollars as its defense budgets and it announced to enhance their defense R&D budget from 6% to 12%–15%, which is equivalent to the United States or European countries in the near future. In Israel, it also spends over 9.7 billion dollars which is over 8.3% of its National GDP ratio MND (2006).

will be very costly to acquire the ability to build defense critical technologies and weapon systems for a self-defense capability in the Republic of Korea.

Defense Acquisition Program Administration (DAPA) of the Republic of Korea was established in 2006, and it has emphasized the importance of independent defense technology and weapon system R&D as one of its main policies (DAPA, 2006). Especially, defense critical technologies which require technology accumulation in long-term periods and crucial fields with independent technical capabilities should be the focus of independent R&D for national security. Therefore, it is more crucial to attain consistent and secure defense R&D budgets to develop defense critical technologies for the reinforcement of national defense strength in the near term.

Given the importance of defense R&D for its technology and weapon systems, more logical and objective valuation of R&D projects is also a crucial instrument in deciding whether to invest or not for making sound choices and budget allocation in defense acquisition fields. However, traditional cost approaches based on net present value (NPV) analysis is criticized due to the lack of consideration of defense R&D characteristics including its irreversibility, uncertainty, and managerial flexibility (Park, 2004). There is a concern that a classic cost–benefit optimization will be a guiding principle for future uncertainties (Johnson *et al.*, 2003). The NPV rule and other discounted cash flow (DCF) techniques for capital budgeting seem to be inappropriate to build the portfolio of research projects as they are in favor of short-term projects, over the long-term and address relatively uncertain projects (Pennings and Lint, 1997).

With this reasoning, this chapter presents a more appropriate valuation model based on the real option approach. Much of the value of R&D is embedded in options (Boer, 1999). This chapter presents the eightfold sequential compound option model to valuate defense R&D projects with an illustrative application using a case study in the Republic of Korea. And also it shows the comparison of different values in two valuation models and presents sensitivity analysis in the variation of relevant parameters. In due course, focusing on the importance of the internal option values in the defense R&D process, it stresses on the application of real options theory not only in a valuation method but also in a new way of thinking for defense R&D projects in the future.

This chapter begins by presenting the nature of the process in the defense R&D projects as a series of options, that is, a sequential compound option, then researches previous literature on both the application of real options in the defense fields and the application of the sequential compound real options. Next, it derives a closed-form solution for the eightfold sequential compound option model with the defense adjusted assumptions and parameters. This model is based on Engelen *et al.*'s (2004) outstanding paper using a generalization of Geske's (1979) compound option model. Following this, the chapter applies a real option model by computing the values of relevant data and compares the traditional NPV-based models. Then it compares two different valuation models and shows a sensitivity analysis of the real option value with respect to its major value drivers. Finally, it presents both discussion and conclusions.

Background

There has been much research on the R&D project selection field over many decades (Souder, 1973a, 1973b; Martino, 1980; Wind and Saaty, 1980; Mandakovic and Souder, 1985; Saaty and Shang, 2007). Souder (1972) suggested the general scoring systems methodology for rating and ascertaining the relative degrees of suitability of R&D project selection models. It is still useful for selecting the best practices among projects because of its simplicity, easy understanding, and economicality. Saaty (1980) presented a powerful R&D project selection model, the analytic hierarchy process (AHP), which is used for scaling the weights of the elements in each level of the hierarchy with respect to an element of the next higher level, a matrix of pairwise comparisons of the activities. Gerdsri and Kocaoglu (2007) also presented the application model using AHP to build a strategic framework for technology roadmapping. With its effectiveness, the AHP is widely used even today as a powerful R&D project selection model in the world. And multicriteria decision-making models are quite effective due to the complexity of the decision problems (Pohekar and Ramachandran, 2004). However, these well-known R&D project selection models have not taken care of the flexibility with respect to the change of future environments. And they also have not presented the real values of the R&D projects to the decision makers, either.

Today, based on its effective valuation tool, a real option model is also effective for use as a R&D project selection model (Black and Sholes, 1973; Ceylan and Ford, 2002). In the defense fields, the real option approaches are also stressed today. In the office of force transformation, DoD, Glaros (2003) emphasized the importance of real options with its flexibility under uncertain combat situations. And he mentioned that Planning, Program, and Budgeting System (PPBS) and Five Year Defense Plan (FYDP) are no more valuable due to less consideration of unknown threats and unstable environments. Therefore, the chapter concluded that even in defense fields, the real option is considered a powerful and flexible valuation method for considering uncertainty in future battlefields.

A sequential compound option means that the execution and value of a strategic option depend on another strategic option in sequence. It is that the execution and value of future strategic options depend on previous options in the sequence of execution (Mun, 2002). The R&D investment is the case of a sequential compound real option, leading to the exercise of follow-on discretionary investment opportunities at subsequent stages. Therefore, it is readily used for defense R&D investment cases, which have similar characteristics of sequential compound real options. Such multistage or compound real options are of strategic import to those firms that possess them. They cannot simply be seen as independent investments but instead should be seen as links in a chain of interrelated projects, the earlier of which are prerequisites for the ones to follow (Panayi and Trigeorgis, 1998).

Sequential compound options are options in sequence. Geske (1979) introduced the notion of a compound option and derived an analytical formula for compound options when the firm value follows a geometric Brownian motion (GBM) within the closed-form solution framework. Geske's framework has provided insights for the analysis of real compound or growth options (Geske, 1979).

Many real investments, for example, in R&D in a strategic acquisition or in an information-technology (IT) network, are undertaken not only for their expected directly measurable cash flows, but also for the future growth opportunities they may unlock, for example, in terms of new technological breakthroughs or access to new markets. Therefore, such investments may be seen as compound options, which have to be evaluated not as independent options, but rather as links in a chain of interrelated projects. Geske's (1979) analytical formula is obtained when volatility is constant. However, because of the second-order nature of compound options, their sensitivity to the underlying parameters is magnified, and then these formulas can hardly be used in practice.

The Process of a Defense R&D Project

It is crucial to decide the way of defense acquisition types in the country. This is mainly divided into two types between domestic R&D and foreign acquisition.

The process of defense R&D projects is characterized by a series of multiphases. According to the DAPA (2007), they are composed of two sub processes. One is the defense technology R&D and the other is weapon system R&D. The defense technology R&D process is composed of three phases from basic research through applied research to experimental development. First, basic research is a theoretical or experimental activity of academic institutes for the purpose of acquiring the necessary hypothesis, theory, phenomena, and observable facts to develop defense core technology or related modules. Second, applied research is a transition phase to the solution of defense matters based on the results of basic research. Researchers have to verify the technical validity and applicability in the experimental environments and develop experimental prototypes. Third, experimental development is the last phases of core technology and module development. This would develop core technology and modules, which are crucial as a major functioning factor of weapon systems, and would verify the usability of present weapon systems and the applicability for future weapon systems.

Weapon system R&D is also composed of three phases from conceptual research and search and development to system development. Conceptual research is a research process necessary to develop independent weapon systems or a not yet acquired weapon system among long-term need projects. It is divided by development alternatives and concept design to execute systematic technology research and analysis. The search and development phase does a risk analysis, technical and engineering explanation, and simulations of subsystem or core modules from previous stages. If necessary, they continue to work on the core technology research and build models for confirming the possibilities for transforming later phases. Lastly, system development is to develop the weapon system, ready for mass production, after technical and managerial test and evaluation. After the weapon system R&D, production lines

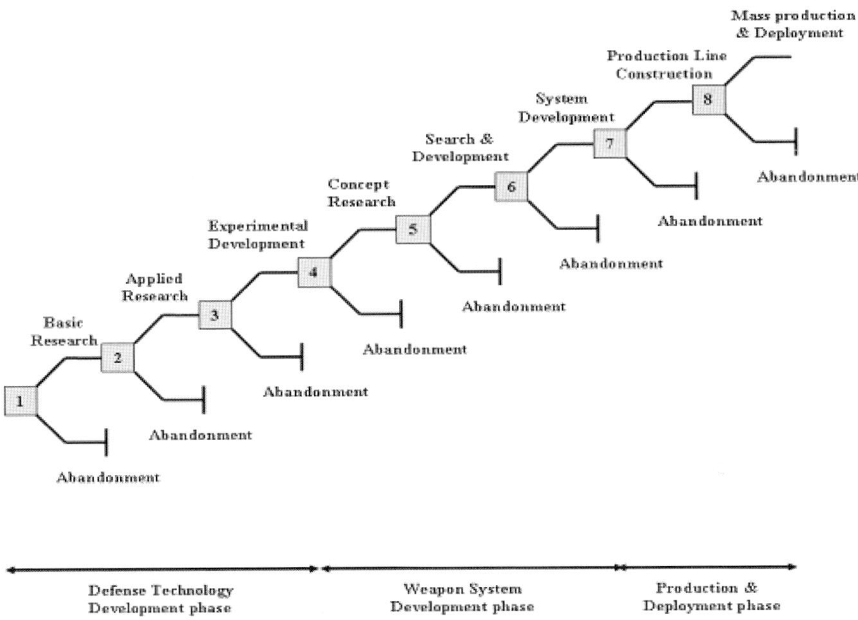

Figure 1: Defense R&D project compound option tree.

are built and made ready for the mass production and deployment to the battle units. Therefore, the whole process of defense technology and weapon system development consists of eight multiphases as shown in Fig. 1.[2]

In summary, the development of a defense R&D project begins with basic research, followed by applied research and

[2]The defense technology R&D and weapon system R&D are identical process. Therefore, they may not perform as one identical process, either. However, based on relevant research, final weapon system is stem from the basic research of necessary defense technology; it is assumed that the defense R&D process is connected as a one whole process from basic research to mass production and deployment (DAPA, 2007).

experimental development. Later, it follows the use of predeveloped defense technology, conceptual research, search and development, and finally the making of the weapon system. After the weapon system development, production line preparation leads to finally finishing the mass production and deployment.

The basic research phase can be seen as an option on the next applied research. If the basic research R&D is successful, then the next applied research R&D is able to proceed; otherwise, the applied research is not executed. For this reason, the defense decision makers have the right, not the obligation, to continue the next phase of the project based on the previous stages of investment for defense R&D projects. This is analogous to the call option of the sequential multiphased compound option. Indeed, the basic research phase itself is a call option on the next applied research phase. This kind of chain reaction is continued until the last mass production phase. As shown in Fig. 1, the process of a defense R&D project is considered as the eightfold sequential compound option.

In the next section, the chapter presents two valuation approaches. One is the basic NPV-based valuation approach, and the other is the eightfold sequential compound option valuation model based on the application of Engelen *et al.*'s, (2004) effort.

The Application of Real Options Theory in the Defense R&D Project: An Eightfold Sequential Compound Option Model

To apply an n-fold sequential compound option, the chapter needs necessary assumptions for the defense R&D project. Based on the assumptions of Black and Scholes (1973) and Engelen *et al.* (2004), it presents adjusted assumptions as follows.

First, the changes in current value of underlying assets in defense R&D projects have unknown values for future uncertainty. Therefore, it is assumed that the value follows a continuous time stochastic process, which is GBM. The mean and the standard deviation are followed by a Markov process, independent increments, and a log-normally distribution. It is also assumed that the rate of return of defense R&D project is followed by a normal distribution. Second, all defense R&D projects are tradeable unlimited. Third, the investors are unsatiated, the security markets are perfect and competitive, unrestricted short sales with full use of proceeds are allowed, the risk-free interest rate is known and constant, the trading is continuously on time, and the firm has no payouts. Fourth, the short-term risk-free interest rate is known to a positive value. Fifth, no transaction costs related to defense R&D projects are possible to import or export with a risk-free interest rate. Last, the dividends are considered as a convenience yield after the development of the R&D project.

It is also necessary to the characteristics of related parameters to valuate the sequential compound real option in a defense R&D project. First, the current value of the underlying asset (V) is the current value of future cash flows incurring from the projects of a defense R&D project. Second, the exercise price for the compound call option (K_i) is the investment costs of each phase in the development of a defense R&D project. Also, the sum of each exercise price is the total investment costs of the defense R&D project. Third, maturity date (*ti*) means the necessary periods of each phase to execute a defense R&D project. Total maturity date (T) is also the sum of each maturity date phase of the defense R&D project. Fourth, risk-free interest rate (r) is basically used as the interest rate of 3-year maturity government bonds with the percentage stated per annum. Fifth, variance (σ^2) represents the variability of defense R&D projects and may be different in accordance with the different characteristics of various

Table 1: Real option parameters for the defense R&D project.

Parameters	Notations	Explanations
Compound call option	C	Compound call option value
Current value of the underlying asset	V	Current value of the defense R&D project
Exercise price	K_i	Defense R&D Investment costs The exercise price for the compound call option C_i
Maturity date	t_i	Maturity date for the compound call option C_i
Volatility	σ^2	Instantaneous variance of the returns on the defense R&D project
Risk-free interest rate	r	Interest rate of 3-year maturity government bonds with the percentage per annum
Cumulative normal dist'n functionon normal ditr'n	$N(d)$	Cumulative normal distribution function
Correlation matrix	A	Correlation matrix

defense R&D projects. It is also presented as a percentage per annum. Finally, cumulative probability distribution (CPD) function of normal distribution (N_i) actually means a cumulative normal distribution function. Table 1 shows the real option parameters and their explanations of the defense R&D project.

With the adjusted assumptions and parameters of the defense R&D project, the chapter presents the eightfold sequential compound option model with the variation of Engelen *et al.* (2004)'s formula as shown in Eq. (1).

$$C_1(V,t) = VN_8(a_1,a_2,\ldots,a_8,A^8) - \sum_{m-1}^{8} K_m e^{-r(t_n-t)} N_m(b_1,b_2,\ldots b_m,A^m)$$

with

$$b_i = \frac{\ln(V/\overline{V}_i) + (r-(1/2)\sigma^2)(t_1-t)}{\sigma\sqrt{t_i - t}}$$

$$i = 1,2,\ldots,8$$

$$a_i = b_i + \sigma\sqrt{t_i - t}, \qquad i = 1,2,\ldots,8 \qquad (1)$$

$$\rho_i = \sqrt{\frac{t_i - t}{t_j - t}} \qquad i < j$$

$$\overline{V}_i = \text{the solution of } C_{i+1}(V,t_i), K_i, \rho_i = 1,2,\ldots,7$$

$$\overline{V}_n = K_n$$

$$A^1 = (a_g^b)_{i,j-1,\ldots 3.} \quad \text{with} \begin{cases} a_{ii} = 1 \\ a_{ij} = a_{ji} = \rho_{jj} \end{cases} \qquad i < j$$

Equation (1) proposes a closed-form solution for the eight-fold compound option. With the previous closed-form solution, it presents the internal option values of a defense R&D project and is able to support more appropriate decision-making tools for defense decision makers whether to invest or not for their defense R&D projects.

An Illustrative Application of the Real Option Model to the Defense R&D Project

According to the MND in the Republic of Korea, the Joint Chiefs of Staff (JCS) announced that the most important needs for future weapon systems was the X project[3] for national security in 2010. There are two options to choose: one is the domestic

[3] Due to the security reasons, it is presented by the X project.

R&D acquisition, and the other is the acquisition of the weapon system from abroad. To choose the best alternative, MND hopes to know the effectiveness and values of the X project. If domestic R&D is to be chosen as the winner, it should prove its potential value.

Here, the chapter presents two valuation methods which include traditional NPV and sequential compound option models. Before proceeding with the real option values of the defense R&D project, it is appropriate to provide a reference benchmark for the results by calculating the traditional NPV values shown in Eq. (2).

$$NPV = \sum PV \text{ of Benefits} - PV \text{ Investment Cost}$$
$$= \sum_{t-1}^{T} \frac{FCF_t}{(1+\text{WACC})^t} - \sum_{t-1}^{T} \frac{I_t}{(1+r_f)^t} \quad (2)$$

This presents the eightfold sequential compound option model to compute the actual value of the X project. As shown in the previous section, the whole development process of a defense R&D project is characterized as the eightfold sequential compound call option. An illustrative application is as follows.

First, the different maturity dates (t_i with $i = 1,2,...,8$) for the compound call options (C_i with $i = 1,2,...,8$) are presented in the approximation for the basic research (1 year), applied research (3 year), experimental development (2 year), concept design (1 year), search and development (2 year), system development (2 year), production line construction (1 year), and finally mass production and deployment (1 year). Therefore, it will take approximately 13 years of total lead time for the X project development. The exercise prices (K_i) are the R&D investment costs at each phase. The estimation of the present value of the mass production and deployment phase (V) is 130 billion won, while the investment cost (I) is estimated to be 50 billion won. The risk-free interest rate is the rate on a 3-year government bond, Bank of Korea, approximately 5%. And the estimation of the volatility (σ^2) is assumed to be 50% due to the high risk and uncertainty for the development. Table 2 shows the necessary data for the calculation of compound option values of the project.

After the use of GAUSS™ program with the data, the NPV value is negative (−10.34 B Won). Due to the incredible amount of initial investments and negative NPV value of the project valuation, there is an inability to begin the X project in the end. On the contrary, when it presents the eightfold sequential compound option approach, the option value is truly positive (29.09 B won). The comparison of the two different values is listed in Table 3.

Table 2: Description of the values of the eightfold compound option. (Unit: Year, B Won, %)

Phase	t_i	K_i	V	I	σ^2	r
I	1	1				
II	3	8.8				
III	2	16.2				
IV	1	1				
V	2	34.1				
VI	2	34.1				
VII	1	10				
VIII	1		130	50		
Total	13	105.2			50	5

Table 3: Comparisons of two values for the defense R&D projects (Unit: B Won).

Model	Final Value	Final Decision
NPV	−10.34	Rejected
Sequential Compound option	29.09	Approved

Therefore, the R&D project could be launched with the positive values from the compound option model.

Conclusions

The result shows that the defense decision maker should take care of real option values for decisions on R&D projects. In the defense acquisition field today, he could ignore the importance of future option values and tend to decide his important decision based on NPVs.

As the trend of complexity, high cost, and fast development of weapon systems is going to increase today, it is hard to decide the R&D project based only on the current NPVs. And even Glaros (2003) emphasized the importance of real option values with its great advantages of flexibility under uncertain situations. The result could direct both the importance of the real option value but also the option mindsets when he decides his important decision in the near future.

Compared to the previous methods of R&D project selection including scoring method, AHP, and others, the real option model could give the decision maker a more flexible way of making his decision and evaluating the importance of future values of high defense technology according to the process of the defense R&D project. The real option model should contribute to selecting the high-technology project in the major consideration of its option values. Also it can be applied to selecting dual use technology, technology transfer, and technology financing with its theoretical backgrounds in the near future.

In conclusion, in the major research of defense R&D characteristics including its irreversibility, high risks and uncertainty, and manager's flexibility, the chapter concludes that the valuation methods of the defense R&D projects are appropriate for using the multiphase compound option, rather than the traditional NPV method. The real option mindset could give the defense decision makers a more valuable model for their important decisions on defense R&D projects.

References

Black, F. and Scholes, M. S. (1973). The Pricing of Options and Corporate Liabilities, *Journal of Political Economy*, **83**: 637–659.

Boer, F. P. (1999). *The Valuation of Technology*. New York, NY: John Wiley & Sons.

Brealey, R. and Myers, S. (1991). *Principles of Corporate Finance*, 5th Ed. New York, NY: Irwin McGraw-Hill.

Ceylan, B. K. and Ford, D. N. (2002). Using Options to Manage Dynamic Uncertainty in Acquisition Projects, *Acquisition Review Quarterly*, 241–258.

Cho, J. W. (1995). Empirical Research on B-S OPM: Comparison of the Merton Model, *Master Thesis of the Seoul National University*.

DAPA (2006). *Defense Projects Law*. Seoul: DAPA.

DAPA (2007). *Defense Project Management Regulations*. Seoul: DAPA.

Engelen *et al*. (2004). The Valuation of a NDA Using a 6 Fold Compound Option, *Research Policy*, 41–51.

Gerdsri, N. and Kocaoglu, D. F. (2007). Applying the Analytic Hierarchy Process (AHP) to Build a Strategic Framework for Technology Roadmapping, *Mathematical and Computer Modeling*, **46**: 1071–1080.

Geske, R. (1979). The Valuation of Compound Options, *Journal of Financial Economics*, **7**: 63–81.

Glaros, G. (2003). *Real Options for Defense*. Arlington, VA: U.S. Department of Defense.

Hwang, Y. (1990). Research on the Pareto Efficiency of the Stock Markets and the Contribution of Real Options, *Master Thesis of the Seoul National University.*

Jang, Y. (1990) Research on the Price Decision Models of the Future Options, *Master Thesis of the Seoul National University.*

Johnson, S., Martin, L., and Treverton, G. (2003). *New Challenges, New Tools for Defense Decisionmaking*. Santa Monica, CA: RAND Corporation.

Mandakovic, T. and Souder, W. E. (1985). An Interactive Decomposable Heuristic for Project Selection, *Management Science*, **31**(10): 1257–1271.

Martino, J. P. (1980). Technological Forecasting — An Overview, *Management Science*, **26**(1): 28–33.

Ministry of National Defense. (2006). *Defense White Paper*. Seoul: MND.

Mun, J. (2002). *Real Options Analysis: Tools and Techniques for Valuing Strategic Investments and Decisions*. New York, NY: John Wiley & Sons.

Panayi, S. and Trigeorgis, L. (1998). Multi-Stage Real Options: The Cases of Information Technology Infrastructure and International Bank Expansion, *Quarterly Review of Economics and Finance*, **38**: 675–692.

Park, J. Y. (2004). The Study on Decision Factors of Optimal Timing of New Weapon System Introduction as a New Investment Opportunity from Real Option Approach, *Master Thesis of the Seoul National University.*

Pennings, E. G. and Lint, O. (1997). The Option Value of Advanced R&D, *European Journal of Operational Research*, **103**: 83–94.

Pohekar, S. D. and Ramachandran, M. (2004). Application of Multi-Criteria Decision Making to Sustainable Energy Planning—A Review, *Renewable and Sustainable Energy Reviews*, **8**: 365–381.

Saaty, T. L. (1980). *The Analytic Hierarchy Process*. New York, NY: McGraw-Hill.

Saaty, T. L. and Shang, J. S. (2007). Group Decision-Making: Head-Count Versus Intensity of Preference, *Socio-Economic Planning Sciences*, 22–37.

Souder, W. E. (1972). A Scoring Methodology for Assessing the Suitability of Management Science Models, *Management Science*, **18**(10): B-526–B-543.

Souder, W. E. (1973a). Analytical Effectiveness of Mathematical Models for R&D Project Selection, *Management Science*, **19**(8): 907–923.

Souder, W. E. (1973b). Utility and Perceived Acceptability of R&D Project Selection Models, *Management Science*, **19**(12): 1384–1394.

Trigeorgis, L. (1996). *Real Options; Managerial Flexibility and Strategy in Resource Allocation.* Boston, MA: MIT Press.

Wind, Y. and Saaty, T. L. (1980). Marketing Applications of the Analytic Hierarchy Process, *Management Science*, **26**(7): 641–658.

CHAPTER 8

R&D Portfolio Management: Integrated Technology Roadmapping Tool to Aid the Decision-Making of R&D Investments

Husam Arman, Nabil Gindy,† Mohammad Kabli‡ & Shirley Cavin†*

Aligning R&D projects to the strategic objectives of every advanced technology firm is one of the important issues for the decision-makers in these firms. Several criteria's are involved when R&D proposals are evaluated and the projects are selected. However, it is usually the criteria of financial returns, technology alignment, and probability of success that has a greater weight of importance. Technological innovation has been an important driver for competitive success in many industries (Schilling, 2012). For advanced-technology firms, the shorter product life cycles, and global competitions have increased the importance of R&D and consequently the interest in R&D portfolio selection methods. Robust assessment of innovation potential of technologies is essential before investment decision (Dereli and Altun, 2013). At the present time, investments in R&D project if not done properly could lead to wastage of large amount of

*Strategic Planning and Evaluation Office, Kuwait Institute for Scientific Research (KISR), Kuwait.

†Department of Mechanical, Materials and Manufacturing Engineering, University of Nottingham, Nottingham, UK.

‡Department of Industrial Engineering, King Abdulaziz University, Jeddah, Saudi Arabia.

resources or even bring down the enterprise (Heidenberger and Stummer, 1999).

The success of the R&D portfolio selection methods depends on the appropriateness of the inputs to develop the right R&D strategies. An important aspect is the integration and the alignment between business and the R&D strategy. In scientific literature and management practice, there is a considerable agreement on the importance of accurate and early anticipation of future needs and technological developments (Reger, 2001). Therefore, an integrated technology planning process is required to contribute to the development of the final R&D portfolio.

The portfolio selection methods described in literature have many different forms that vary between subjective judgment of R&D managers and formalized techniques. In benefit measurement methods, projects are ranked based on the amount of benefit they provide and then sequentially selected within overall budget constraints. Examples of benefit measurement methods are analytic hierarchy process (AHP) (Suh *et al.*, 1994), analytic network process (ANP) (Mohanty *et al.*, 2005), q-sort (Souder and Mndakovic, 1986), and economic models (Lint and Pennings, 1998). Mathematical programming methods tend to optimize one or more objective functions subject to constraints of different aspects, for example, budget and technology type (Stummer and Heidenberger, 2003). Under conditions of uncertainty, both decision and game theory take into consideration the possible occurrence and extent of future events and their impact on a company (Ali *et al.*, 1993; Heidenberger, 1996). In order to reach an estimate range of possible outcomes and the probability of different outcomes, projects are simulated using random numbers (Martino, 1995). Heuristic models look for acceptable solutions which are not necessarily optimal (Venkatraman and Venkatraman, 1995), while cognitive emulation methods attempts to build on previous experience in order to establish a model of the actual decision-making process within an organization (Hall and Nauda, 1990).

In reality, R&D portfolio selection methods are not used widely. The methods highlighted previously lack one or more of the following issues (see, for example, Cooper *et al.*, 2001; Martino, 1995; Stummer and Heidenberger, 2003):

- Treatment of multiple, often inter-related, criteria including risk, uncertainty, and nonmonetary aspects.
- Treatment of project inter-relationship with respect both to value contribution and to resource utilization.
- Perceptions held by the R&D managers that the models are unnecessarily difficult to understand and use, and do not engage their experience and knowledge.
- Treatment of the time variant property of data and criteria, and the associated problem of consistency in the research program and the research staff.
- The portfolio reflects the business's strategy.

The model we propose addresses most of the above issues. This is achieved by pairing various concepts represented in an integrated framework to support strategic technology planning as part of corporate strategies. For example, the proposed model is based on a method that uses technology alignment and deals with the risk in addition to the traditional monetary criteria. The inter-relationships between projects can be included using simple dropdown menu which translates into equations as part of the linear programming model. This is only an example of how the system tried to simplify the process where the user need not worry about understanding optimizations algorithms. Finally, the R&D portfolio is derived based on the technology requirements that is developed and aligned with the business strategy.

We present a model that optimizes portfolio decisions, which is based on an integrated technology planning process aimed at aligning enterprise R&D activities to the business objectives through identifying, selecting, prioritizing, and developing technologies to satisfy market and product needs.

The contribution of the proposed model lies in the integration dimension which uses the company's quantitative and qualitative objectives within R&D management in a simple and useful process. Moreover, several issues are considered and expressed in the constraints of the model to represent real situations and applications. However, it is assumed in this specific application that all the R&D projects are new, and that an independent budget is allocated to these projects.

The Integer linear programming (ILP) model has been programmed and embedded as part of dedicated STAR software which captures all the data generated during the implementations of the technology roadmapping exercise. The ILP software utilizes the algorithm that is available as an open source to solve general linear programming problems. Inputs and outputs interfaces were designed and developed to serve specific R&D problem.

A Requirement for an Integrated Technology Planning Approach

Technology roadmapping is often used to support strategic and long-term planning. It links market, product, and technology to justify company's investments strategies (Albright and Kappel, 2003). Companies apply technology roadmapping primarily to ascertain what an evolutionary future is likely to offer, and to indicate actions that could be taken to be successful in future (Kappel, 2001). It is carried out when organizations (departments, companies, sectors, countries, or even economic regions) come to a decision point, for example, prior to an annual allocation of funds or when a crisis due to a rapid loss of market share. Phaal *et al.* (2004) emphasize the purpose of the Technology Road Map (TRM) as a communication tool between the different functions inside the firm.

Several contributions toward integrating technology roadmapping and business needs to support planning and for technology as part of the firm general business planning (Petricka and Echols, 2004; Phaal *et al.*, 2003, 2004; Oliveira and Rozenfeld, 2010), yet few attempts such as (Loyarte *et al.*, 2015) to utilize an integrated technology roadmapping that can lead to an optimized portfolio is still needed.

STAR goes beyond technology roadmapping, since the key enhancement of STAR is not just the integration and the visualization of the business and technology strategy. These functions drive the actions and the activities to be followed by proposing, assessing, and selecting the right R&D investments. STAR promotes that technology roadmapping should be an integral part of the management of R&D and technology. Technology roadmapping aim to influence, drive, and envision the process of R&D resource allocations considering the strategic alignment with the support of a simple and reliable computerized optimization technique embedded in the STAR software.

Kappel (2001) states that an essential perspective on technology roadmapping is its use as a method of strategic technology planning. However, the usefulness of technology roadmapping is realized once actions are followed and decisions are made based on the business needs. STAR addresses these gaps and goes beyond roadmapping through aligning and linking technology strategy to the right market and business strategy.

An Overview of STAR Methodology

The overall STAR framework integrates and aligns five main trajectories: business, market, product, technology, and research and development each of which is made of several processes. Each process is designed to integrate several steps. The overall objective is to optimize benefits from technology development projects and their contribution towards achieving enterprise goals. The outputs of STAR are a set of STAR maps designed to aid communication of the results of the roadmapping process. The maps provide a consensus-based transparent view of business drivers, market and product requirements, competitive position, and the technology landscape and R&D project portfolio.

The STAR process can be broken down into major steps (activities). These activities are accomplished using workshops approach supported with dedicated and integrated software developed as part of STAR methodology at the University of Nottingham (Gindy *et al.*, 2008).

The STAR framework is based on closely coupling several techniques and methodologies to provide an integrated framework for enterprise technology requirement planning and technology investment strategies (see Fig. 1 for the generic framework).

One of the major outputs of STAR is a portfolio of R&D projects that optimize the allocation of resources to the projects with high impact on technology development and projects that meet enterprise business needs.

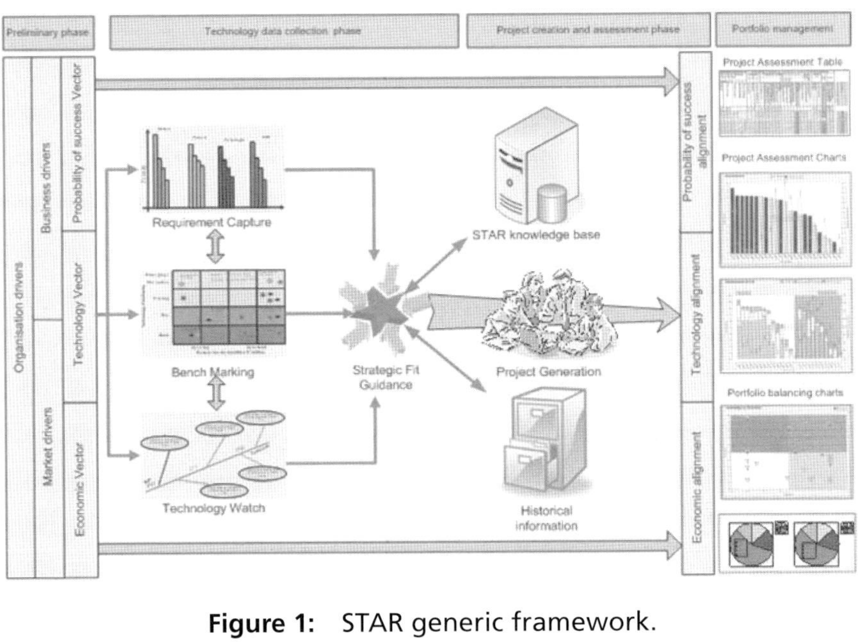

Figure 1: STAR generic framework.

Preliminary Phase

The preliminary phase takes place prior to the technology roadmapping exercise, and is responsible for ensuring that preparation work has been carried out appropriately.

For a full database-base implementation of STAR, this preparation will include populating the database with products, components, processes, and technologies which is a software-assisted process. The preparation work should involve several teams with appropriate level of knowledge regarding the enterprise organizational structure, markets, current and future products, product life cycles, business drivers, supply chain, make and buy strategies, and the manufacturing technologies used in product manufacture. These teams will provide and facilitate populating and operating the knowledge management system which is controlled by hierarchical user privileges.

Technology Data Collection Phase

The core element of any technology management process is the task of deciding about the actions which usually materialize R&D projects or R&D programs. The key process in STAR is the development of a strategic fit guidance (see Fig. 1). This is a statement based on the company's current technological requirements (requirements capture process), assessing company's technological competitive position (benchmarking process), and the future technological landscape (technology watch process) to guide the strategic alignment of the enterprise (Arman and Foden, 2010; Hodgson *et al.*, 2008).

Project Creation and Assessment Phase

At this stage, all the information and outputs are ready to be conveyed to the decision-maker in the enterprise. Several figures and charts are designed to aid in understanding the inputs/outputs and the consequences of decision-making parameters.

The project creation is driven by strategic fit guidance which represents the target technology development domain most aligned to enterprise circumstances that technology development projects should strive to achieve. This statement is determined through a consensus agreement by company experts, who express their preferences regarding the nature of the technologies the enterprise should invest in, and the thresholds of the economic and associated risks.

STAR Software

Many companies use general platforms such as spreadsheets and graphical software but little commercial software dedicated to this purpose are available such as Accolade Vision Strategist (www.sopheon.com) and TeMAP (www.bmt-softwaresolutions.com). These systems are very useful for managing and monitoring the company's projects. The systems make it easy to share product and technology roadmapping strategies in a secure mode with suppliers and partners. However, systematic and linked technology roadmapping processes from business to R&D strategy and optimized decision-making are not enhanced in these systems.

STAR software is an application that is based on the STAR methodology and technology roadmapping processes. The system is composed of an integrated knowledge base and user-friendly software. The STAR knowledge base integrates all aspects related to a generic technology roadmapping process. It is designed to support the STAR software in every stage, providing relevant information to carry out each exercise as part of STAR methodology. The knowledge base includes a market knowledge base, a product knowledge base, a technology knowledge base, and an R&D knowledge base. The STAR user-friendly software allows the user to conduct a series of exercises that are part of the STAR methodology and provides valuable outputs that aims to support the decision making process, such as market, product and technology assessments, project evaluation, and optimum portfolio generation. The software integrates a set of exercises each of them targeting an important aspect of the strategies considered in the STAR methodology, as illustrated in section "Portfolio Balancing and Optimization Tool as Part of STAR Methodology".

Portfolio Balancing and Optimization Tool as Part of STAR Methodology

The optimization tool enhanced in the integral system comprises the principal engine to produce an optimized portfolio of projects to meet the business and market requirements of the organization. It is important that an appropriately balanced portfolio is achieved in order to ensure that:

- An effective mix of technologies is developed for insertion into the enterprise.
- An appropriate balance is achieved between short-term and long-term projects, diversification of risks.
- Utilization of human, plant, and other resources within enterprise capacity.

Eventually, the task of ensuring an effective portfolio will require the experience and judgment of decision-makers to set the balancing rules and make sure the chosen portfolio is strategically aligned with the company's business objectives. A straightforward process of portfolio balancing and optimization is presented in Fig. 2, where a screening process is applied before and after the optimizing process, using ILP model, to confirm balanced portfolio aligned with the company's preferences.

Mainly three criteria are utilized to assess projects attractiveness. The first is the economic dimension which takes into account the estimated financial returns considering the financial investment required to carry out the project. The second dimension is the technology alignment which measures quantitatively the cumulative impact of the R&D project technology

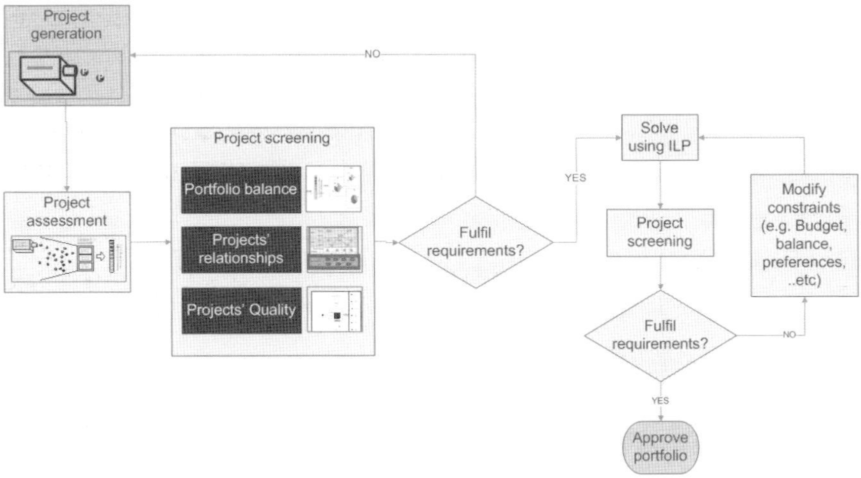

Figure 2: Process map of portfolio balancing.

pursued on the enterprise goals and its competitive position. It is based on the information collected during the requirement capture, benchmarking and technology watch steps (see section "Objective of the Model"). The third dimension in the criteria of project assessment is the synergistic alignment. This considers all the risks and opportunities associated with projects. Although some of these factors are qualitative and difficult to estimate, they can have, in many cases, vital impact on the chances of a project achieving its objectives.

The screening process represents a sanity check and it includes several graphical outputs (STAR maps) which are provided as part of STAR software to assist in this activity (see Fig. 3). The main purpose of the STAR maps is to ensure the following:

1. The expected portfolio is balanced based on the company's definition (e.g., project size, time horizon, and risk). For example a maximum of 20% of company's budget is allocated for small or high-risk projects.
2. The relationships between projects (interdependency) are checked and visualized for example using a typical technology roadmapping chart.

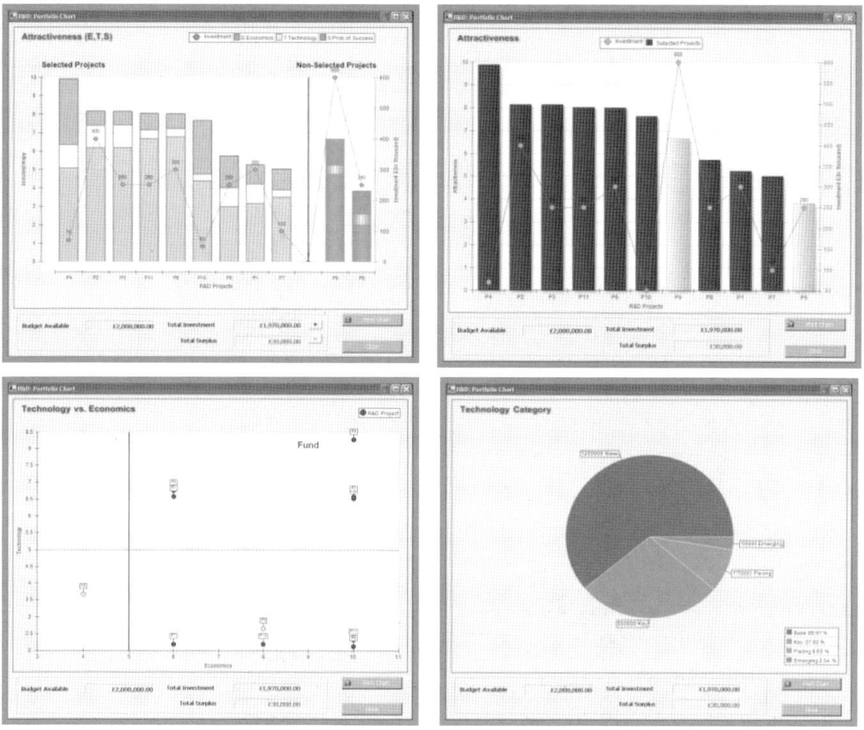

Figure 3: Examples of STAR maps.

3. The quality of projects is evaluated as a collection (i.e., potential portfolio) by looking at the distribution of the projects in relation to economic, technology, and probability of success.

The optimum portfolio represents the most attractive projects as a combination that fulfils quantitative and qualitative objectives. The ILP model is based on the projects' contribution toward company's business and market requirements utilizing the three criteria used in STAR assessment phase.

The ILP model is embedded in the STAR software which provides the facility to use the optimization technique in a user friendly, graphical and customized interface, and dedicated to and integrated within the R&D portfolio management section of STAR methodology. The final portfolio will then be visualized again to be checked by the decision-makers to ensure the appropriateness and the adequateness of this portfolio in reflecting the preference and the needs identified explicitly and earlier in STAR process. The approved portfolio with all the evidences (e.g., STAR maps) generated will be stored in the STAR database.

ILP Model Formulation

In this model, the decision variables in the binary IPL can only take the values one "1" or zero "0" where "1" means the project has been selected by the model while "0" the project has not been selected (the mathematical programming formulation is included in the Appendix).

Objective of the Model

The objective of this model is to maximize the tangible and intangible benefits of R&D projects proposed at a company level by selecting the best combination of the projects proposed (i.e., an optimal portfolio). This is achieved by maximizing the overall attractiveness of the proposed projects considering various issues and constraints. Each project has attractiveness value based on its contribution and alignment to the company's strategic objectives and drivers. The coefficients associated with these variables are the attractiveness, which has been explained in the above discussion. Based on this, a simple and straightforward equation was formulated

$$\text{Max}\left(\sum_{i=1}^{k} A_i \times X_i\right) \tag{1}$$

where,

X — 0–1 variable, If X_i is 1 then project i will be selected. If it is 0, investment will not be made,

A — Project's attractiveness,

K — Total number of projects.

Constraints of the Model

In portfolio optimization problem, constraints are restrictions and issues that are utilized to manage how the money (i.e., the budget) can be allocated to produce a balanced portfolio and therefore, the objective of the model are subjected to several groups of constraints. The following are examples of relevant issues that have been deduced based on the literature in the field of R&D management and feedback from the industry and partners.

Budget

Two issues are considered under the budget constraint. They are the overall R&D budget constraint as well as the distribution of the budget by category. An example of the category considered while testing the model, are the "futurity" (i.e., timeline) and the size of the projects. The size of a project is based on the amount of investment while current/future is based on the product/process targeted if it is a current or new technology. This will ensure a balanced portfolio that reflects company's requirements and perspectives.

For the overall budget constraint

$$\sum_{i=1}^{k} I_i \times X_i \leq \text{Total Budget} \tag{2}$$

where,

I — The financial investment required to carry out the project,
K — Total number of projects.

However, for budget by category constrain:

$$\sum_{i=1}^{m} I_i \times X_i \leq \text{Total Budget per category} \tag{3}$$

where,

I — The financial investment required to carry out the project,
M — Total number of projects per category.

The relationship between projects

In reality, some relationships may exist between projects, and therefore restrictions are placed on the number of projects used. For example, maximum number of projects in this area (e.g., machining) is 2 can be translated into (X1 + X2 + X3 + X4 < = 2).

Resources

Different types of resources are relevant when considering R&D projects selection. Important example is the human resources where limitation on manpower affects the investment decision of the R&D projects. Other examples include hardware such as machines and equipment. This type of constraint is illustrated as follows:

$$\sum_{i=1}^{k} R_i \times X_i \leq \text{Total of } R \tag{4}$$

where,

R-Resources unit (e.g., manpower),
k–Total number of projects

Qualitative issues

Qualitative issues are open to almost any issue that is required to reflect company's preference in relation to nontangible factors. In this ILP model most of these issues should have been assessed in STAR process such as opportunities and risks. However, other constraints can also be represented without difficulty to reflect other requirements and restrictions. Examples of these issues are project impact on firm's image, political and environmental changes as long as they are quantitatively scored as part of the project's evaluations.

Case Study: Optimization Model Application as Part of STAR

STAR system was originally designed primarily for optimizing the R&D investments in the manufacturing processes of large and high-technology companies (Gindy *et al.*, 2006). The software-supported optimization tool as part of STAR has been tested in medium and a large technology-based companies. However, the detailed case presented on this section if for a large firm in aerospace.

The implementation of STAR methodology has been completed by planning a set of sessions which included interviews and workshops to acquire and analyze the collected data. The organization of the sessions is illustrated in Fig. 4, which explains the management of the case study application.

The planning stage was a set of initial meetings with the company (senior staff) to decide about the objectives of the exercise and forming the teams. In the preliminary stage, technical data was collected including market sectors and their associated products and technologies for demonstration purposes. In the first workshop several exercises were completed and a strategic fit guidance was issues based on all the results. Projects were selected considering the strategic fit guidance to be evaluated in the second workshop which aimed to generate an optimum portfolio. Each workshop consumed approximately

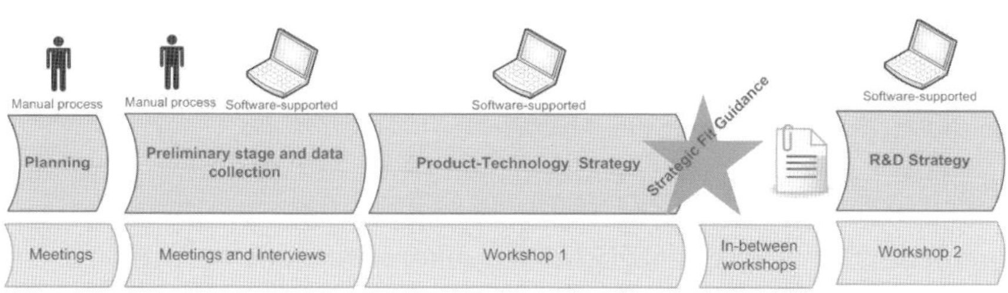

Figure 4: Flowchart describes the case study organization.

half a day. Part of the feedbacks on the workshops was formally captured using an evaluation questionnaire.

The implementation of the STAR was executed as part of a dedicated project inside the company. Senior management in the company appointed a team to support all the organizational issues and logistics in the project. This set in motion a successful implementation of the project. The interviews and the workshops were conducted by the University of Nottingham researchers during on-site companies' visits. The STAR software was provided and demonstrated by the researchers while IT hardware facilities were provided by the companies.

Due to confidentiality agreement signed with the company, it is very difficult to demonstrate using the real data. However, some results can be presented taking out the real names, and so forth. The actual key exercise was to test the STAR approach against the current system utilized in the company to raise the issues of concern and enrich the discussion later on. To do this, the collection of last year projects already selected based on the company's methodology was used again and evaluated based on the STAR criteria. However, all the judgments and inputs were by the same people involved in R&D and technology management.

In this exercise, 46 projects were competing for a fixed budget. The optimizations model considered all the assessment criteria and preferences selected by the decision maker. At the end of the exercise, the optimization model chose 11 projects from the whole list and ensured that the budget requirement is met accurately. Figure 5 depicts the actual final portfolio selected, showing a simplified rationale in an easy to understand graphical representation.

The exercise has considered one market segment and several products and technologies linked to these products. All the proposed projects tested in this exercise were related to the selected market segment for demonstration purposes. Each project was dealt with as a proposal including all the required information in terms of investment, technology contribution, and the probability of success.

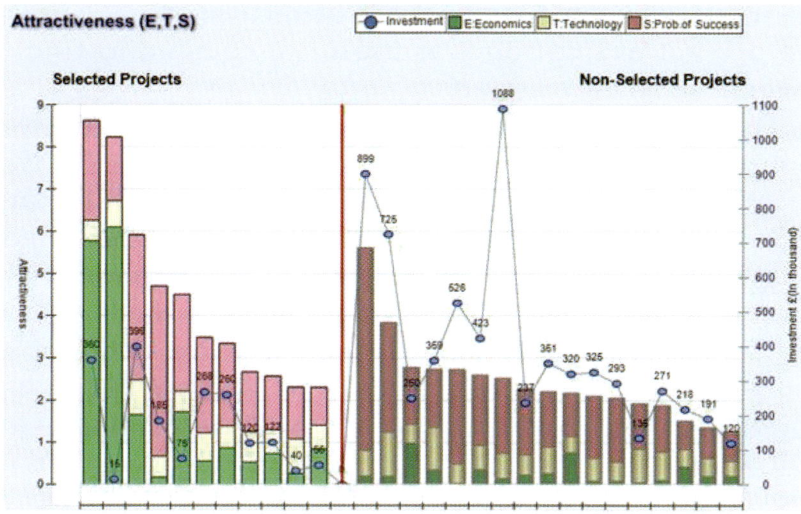

Figure 5: The selected projects using the optimization model.

The STAR software was used to input all the above information and facilitate the analysis process in the workshops. Moreover, the optimization part was supported with friendly screens where decision-makers could see the simplicity of how it works without the need to be familiar with linear programming. This is because all system elements were designed specifically for this purpose compared to other commercial general linear programming software.

After conducting the exercise, another session was organized to investigate and compare the results recorded with the already selected projects using the firm's methodology. As expected, there were some differences in the final selection. Apart from the discrepancy due to the different approaches, several issues were cleared when the company's team commented on the various qualitative, latent, and political issues linked to success or failure of projects to be selected in the final portfolio. However, when the two portfolios were compared, it was easy to trace back the success/failure of any project utilizing STAR software while mainly judgment was used to justify the firms' selection. Moreover, the attractiveness value used in the project evaluation was useful in communicating the importance of each project.

The feedback obtained from the questionnaire on the workshop organization was valuable. The support of the software was widely appreciated due to the flexibility and integration within other STAR functions. The results also showed that industrialists and practitioners are not interested in a range of independent tools in business strategy and technology management which sometimes diminish the simplicity of the real problem. Therefore, it was reasonably received that business and technology management activities should be aligned utilizing an integrated framework to ensure investments are attractive and resources are allocated optimally using a user-friendly system.

Conclusions

The accelerating rate of technology change is increasing the risks that technology orientated companies face. There is a need for decision-making tools to alleviate these risks by aiding technology mangers to invest timely in the right technologies. The methodology outlined contributed to the integration of technology management methods with the support of IPL technique to select manufacturing technology R&D projects aligned with company's business and market requirements.

Optimization techniques proved to be useful if simplified and integrated with methods that allow decision-maker to be involved and responsive to several dimensions relevant to R&D management. The evidences of implementing the proposed model as part of STAR methodology showed advantages and improvements in the outputs comparing to other techniques used in the companies.

It was demonstrated that by using the optimization tool that incorporates the STAR concept of integrating technology roadmapping and business strategy, a more effective and functional strategic technology planning could be developed.

In conclusion, using STAR, can improve the quality of the decision-making process, ensuring R&D investments are strategically aligned with the business requirements, and more practical strategies are realized.

Future work includes further implementation in different sectors, which would give profound insights on advantages and limitations of the methodology which has been tested in limited sectors. A comparative study is being carried out to test the results obtained from applying STAR, including the optimization tool, and the results of the R&D project portfolio obtained by applying the current company practices. Further improvements and advancements to the STAR software are being implemented considering the feedbacks from the case studies.

References

Albright, R. E. and Kappel, T. A. (2003). Roadmapping in the corporation. *Research Technology Management*, **46**(2): 31–40.

Ali, A., Kalwani, M. U. and Kovenock, D. (1993). Product development projects: Pioneering versus incremental innovation strategies. *Management Science*, **39**(3), 255–274.

Arman, H. and Foden, J. (2010). Combining methods in the technology intelligence process: Application in an aerospace manufacturing firm. *R&R Management*, **40**(2): 181–194.

Cooper, R. G. and Edgett, S.J., and Kleinschmidt, E.J. (2001). *Portfolio Management for New Products*, 2nd Ed. New York: Basic Book.

Dereli, T. and Altun, K. (2013). A novel approach for assessment of candidate technologies with respect to their innovation potentials: Quick innovation intelligence process. *Expert Systems with Applications*, **40**(3): 881–891.

Gindy, N. N. Z., Cerit, B., and Hodgson, A. (2006). Technology roadmapping for the next generation manufacturing enterprise. *Journal of Manufacturing Technology Management,* **17**(4): 404–416.

Gindy, N. N. Z., Morcos, M., Cerit, B., and Hodgson, A. (2008). Strategic technology alignment roadmapping STAR® aligning R&D investments with business needs. *International Journal of Computer Integrated Manufacturing,* **21**(8): 957–970.

Hall, D. L. and Nauda, A. (1990). An interactive approach for selection IR&D projects. *IEEE Transactions on Engineering Management,* **37**(2): 126–133.

Heidenberger, K. (1996). Dynamic project selection and funding under risk: A decision tree based MILP approach. *European Journal of Operational Research,* **95**(2): 284–298.

Heidenberger, K. and Stummer, C. (1999). Research and development project selection and resource allocation: A review of quantitative modeling approaches. *International Journal of Management Review,* **1(**2): 197–224.

Hodgson, A., Arman, H., and Gindy, N. N. Z (2008). An intelligent technology watch function for the high technology enterprise. *International Journal of Industrial and Systems Engineering,* **3**(1): 38–52.

Kappel, T. A. (2001). Perspectives on roadmaps: How organizations talk about the future. *Journal of Product Innovation Management,* **18**(1): 36–48.

Lint, O. and Pennings, E. (1998). R&D as an option on market introduction. *Journal of R&D Management,* **28**(4): 279–287.

Loyarte, E., Posada, J., Gaines, S., Rajasekharan, S., Olaizola, I. G., Otaegui, O., Linaza, M. T., Oyarzun, D., Pozo, A., and Marcos, G. (2015). Technology roadmapping (TRM) and strategic alignment for an applied research centre: a case study with methodological contributions, *R&D Management*, **45**(5): 474–486.

Martino, J. P. (1995). *Research and Development Project Selection.* New York: John Wiley & Sons, Inc.

Mohanty, R. P., Agarwal, R. A., Choudhury, A. K., and Tiwari, M. K. (2005). A fuzzy ANP-based approach for R&D project selection: A case study. *Journal of Production Research*, **43**(24): 5199–5216.

Oliveira, M. and Rozenfeld, H. (2010). Integrating technology roadmapping and portfolio management at the front-end of new product development. *Technological Forecasting and Social Change*, **77**(8): 1339–1354.

Petricka, I. J. and Echols, A. E. (2004). Technology roadmapping in review: A tool for making sustainable new product development decisions. *Technological Forecasting and Social Change*, **71**(1–2), 81–100.

Phaal, R, Farrukh, C. J. P., and Probert, D. R. (2003). Technology roadmapping: Linking technology resources into business planning. *International Journal of Technology Management*, **26**(1): 2–19.

Phaal, R., Farrukh, C. J. P., and Probert, D. R. (2004). Technology roadmapping — A planning framework for evolution and revolution. *Technological Forecasting and Social Change*, **71**(1–2): 5–26.

Reger, G. (2001). Technology foresight in companies: From an indicator to a network and process perspective. *Technology Analysis & Strategic Management,* **13**(4): 533–553.

Schilling, M. (2012). *Strategic Management of Technological Innovation.* Noida: Tata McGraw-Hill Education.

Souder, W. E. and Mndakovic, T. (1986). R&D project selection models. *Journal of Resources Management,* **29**(4): 36–42.

Stummer, C. and Heidenberger, K. (2003) Interactive R&D portfolio analysis with project interdependencies and time profiles of multiple objectives. *IEEE Transactions On Engineering Management,* **50**(2): 175–183.

Suh, C. K., Suh, E. H., and Baek, K. C. (1994). Prioritizing telecommunications technologies for long-range R&D planning to the year 2006. *IEEE Transactions On Engineering Management,* **41**(3): 264–275.

Venkatraman, R. and Venkatraman, S. (1995). R&D project selection and scheduling for organizations facing product obsolescence. *Journal of R&D Management,* **25**(1): 57–70.

Appendix — ILP Model Formulation

The objective of this model is to maximize the total attractiveness of the selected portfolio.

Maximize:

$$\sum_{i=1}^{k} A_i \times X_i$$

Subject to:

$$\sum_{i=1}^{k} I_i \times X_i \leq \text{Total Budget}$$

$$\sum_{i=1}^{l} I_i \times X_i \leq \text{Total Budget per category}$$

$$\sum_{i=1}^{k} R_i \times X_i \leq \text{Total of } R$$

where,

Xi — Binary value indicating funding of project *i*,
Ai — Attractiveness of project *i*,
I — The financial investment required to carry out the project,
R — Resources unit (e.g., manpower),
K — Total number of projects.

CHAPTER 9

R&D Portfolio Management: Performance Management

Hakyeon Lee, Moon-Soo Kim,† Soung Ryong Yee‡ & Kyungil Choe§*

With the intensification of global competition for science and technology, R&D has emerged as the key factor in boosting national competitiveness. Many countries have been raising R&D investments for implementation of various national R&D programs and management of research institutes (Lee *et al.*, 1996). One of the most important prerequisites for benefiting from national R&D is an effective and objective system of performance evaluation and management (Nakamura and Tsuji, 2004). Various frameworks and methods have been proposed to evaluate R&D performance, but they have mainly focused on evaluation of short-term R&D outputs despite the fact that there is a time-lag between R&D inputs and outputs (Graves

*The Graduate School of Public and Information Technology, Seoul National University of Science and Technology, 172 Gongreung 2-dong, Nowon-gu, Seoul, 139–176, Republic of Korea.
†Department of Industrial and Management Engineering, Hankuk University of Foreign Studies (HUFS), San 89, Wangsan-ri, Mohyeon-myun, Yongin-si, Kyongki-do 447–791, Republic of Korea.
‡Department of Industrial and Management Engineering, Hankuk University of Foreign Studies (HUFS), San 89, Wangsan-ri, Mohyeon-myun, Yongin-si, Kyongki-do 447–791, Republic of Korea.
§Department of Industrial and Management Engineering, Hankuk University of Foreign Studies (HUFS), San 89, Wangsan-ri, Mohyeon-myun, Yongin-si, Kyongki-do 447–791, Republic of Korea.

and Langowitz, 1996). Long-term R&D outcomes have been neglected so that they have failed to be fully utilized and spread. Thus, it is imperative to mirror long-term R&D outcomes in performance evaluation and utilize them through continuous management and monitoring.

Such needs have been recognized and transformed into policy actions. In Korea, the government enacted "the Act on Performance Evaluation and Management for National R&D Programs" in 2005, and established "A Basic Plan for the Management and Utilization of R&D performance", led by the Ministry of Science and Technology (MOST) in 2006. However, models or frameworks, and methods for implementing the policy properly are unavailable.

In response, this chapter presents a model of R&D Performance Monitoring, Evaluation, and Management System (RPMEMS) for national R&D. The proposed RPMEMS model is targeted to the Korean national research councils, which is in charge of management and evaluation of government research institutes (GRIs) of Korea, and aimed at effectively spreading the outcome of national R&D. First, the conceptual model of RPMES is explained with its scope, structure, role, and process. The RPMES model is based on the logic model used in the planning and evaluation of public programs. In addition, to efficiently utilize limited R&D resources, relative performance based on monitoring results should be measured and corresponding budgets should be allocated to GRIs. For this purpose, two methods are introduced as alternatives for relative comparison among GRIs: analytic hierarchy process (AHP) and data envelopment analysis (DEA), both of which have effectively been used in R&D evaluation (Lee *et al.*, 2008a, b). The AHP proposed by Saaty (1980) is one of the most widely used multiple criteria decision making (MCDM) methods, and DEA is a linear programming model for measuring the relative efficiency of decision-making units (DMUs) with multiple input factors and output factors (Cooper *et al.*, 2000). Since the model of RPMEMS alone cannot make it possible to conduct relative comparisons among GRIs. However, to efficiently utilize limited R&D resources, relative performance based on performance monitoring and evaluation results should be measured and corresponding budgets should be allocated to GRIs accordingly.

Section "Background" reviews the historical background of evaluation of national R&D in Korea and needs for implementing RPMEMS. A brief explanation of the logic model is also given in this section. The RPMEMS model is presented in section "RPMEMS". Section "Methods for Performance Comparison in RPMEMS" proposes methods for relative comparison among GRIs. Section "Conclusions" ends with limitations of this research and future research directions.

Background

Evaluation of National R&D in Korea and Needs for Implementing RPMEMS

With regard to the evaluation of national R&D programs, the current law provides for the need and obligation to monitor and evaluate R&D performance. First, Presidential Ordinance No. 17429 Regulation on the Management of National R&D Programs, and so on issued in December 2001 provides the ground for the monitoring and evaluation, and afterwards the Act on Performance Evaluation for National R&D Programs, and so on and Performance Management enforced in December 2005 provides for specific details on the monitoring and evaluation. In August 2006, based on the act, the National Science and Technology Council (NSTC) formulated a basic plan on the management and utilization of R&D performance, leading to the establishment of a government-wide plan on the management and utilization of national R&D performance results, and to the formulation of a government-wide roadmap for research performance monitoring and evaluation with a view to the open and transferable performance management.

The basic plan on the management and utilization of R&D performance involves the mandated submission of R&D utilization plans at the stages of selection and completion of R&D programs, and its reflection in the evaluation. Notably, heads of major GRIs are required to draft and submit reports on the utilization of research performance every year for a certain period of time following the completion of research. The act and the plan provide that research programs with over certain sizes of costs, or during their development, should be monitored and evaluated, and that any difficulty in utilizing such performance results should be identified along with the provision of appropriate support. To that end, guidelines for target programs, evaluation methods, and so on should be prepared and reflected in the joint management regulations; with regard to the specific selection criteria for target programs, and evaluation time and procedures, relevant ministries are allowed to autonomously perform this work according to the characteristics of their individual programs. Also, the act and the plan provide that, with regard to programs found to undergo difficulty in utilizing their research performance results as a result of monitoring and evaluation, appropriate support should be reinforced. Likewise, the act and the plan reflect the emphasized importance of a systematic management of R&D performance results, and notably aim to maximize the transfer and utilization of research performance results. Thus, the need is raised to introduce a Performance Monitoring, Evaluation, and Management System designed to efficiently support the transfer and evaluation of GRIs' performance results.

Logic Model

The logic model has been widely used to describe and explain the effects of public programs; it is a linear model comprised of an array of logic connections to show how to operate such programs to resolve given problems (Bickman, 1987). With regard to the operation of programs, the logic model can visually display resources, planned activities, targeted changes, and relations between results, thereby allowing the program managers and relevant personnel to share the purpose and details of such programs, and enabling evaluators to induce evaluation tools (Kellogg Foundation, 2004).

As for its advantages, the logic model can first enhance the understanding of diverse aspects of the program, thereby enabling the sharing of ideas, and effective communications. Second, the logic model can help to find important projects geared toward achieving objectives, as well as duplicated or inconsistent factors of the program's factors, thereby providing useful information for designing and improving the program. Third, the logical model can establish balanced measurement and evaluation factors in diverse aspects (McLaughlin and Jordan, 1999).

Diverse logic model types have been introduced and utilized; however, their basic framework does not differ greatly. Figure 1 shows the basic logic model (Wholey, 1983, 1987; McLaughlin and Jordan, 1999). This model, based on delivering the program output to the client, consists of the delivered program in the first half section, with the results from the program shown in the second half section.

The delivered program, which is related to the actual program performance portion, consists of resources, activities, and outputs: "resources" refers to inputs such as the human and material resources required for performing the program;

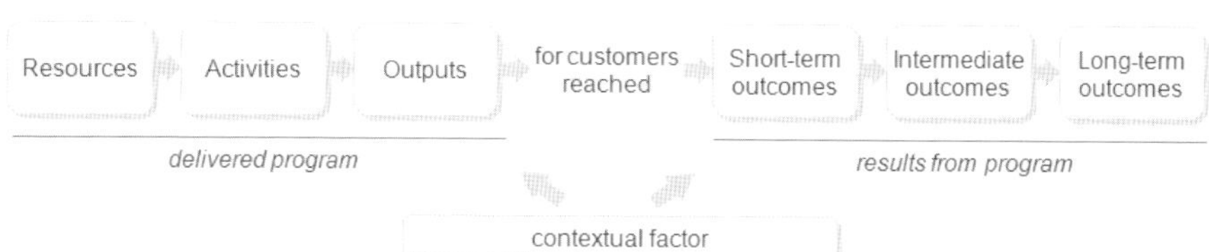

Figure 1: Basic structure of logic model.

"activities" refers to the processes, tools, events, technologies, and acts which are necessary for producing outputs on the basis of the resources; "outputs" refers to the products and services of the program, which are provided to customers. Meanwhile, the effects as a whole that are generated after these have been delivered to the customers are referred to as "the results from the program". The results from the program, according to their creation point in time, are divided into short-term outcomes, intermediate outcomes and long-term outcomes: "short-term outcomes" refers to the changes or benefits that are directly caused by or are closely related to outputs; "intermediate outcomes" refers to the changes or benefits that can be obtained by applying short-term outcomes; and "long-term outcomes" refers to the changes or benefits that are generated from the intermediate outcomes. Finally, "contextual factors" are the external factors that have positive or negative effects on the success and failure of the program.

Some adjustments to the basic logic model have been proposed according to the purposes of the public program and its application targets such as Ruegg and Feller (2003), Kellogg Foundation (2004), and Kocher *et al*. (2006). Notably, many attempts have been made to apply the logic model to public R&D programs. Program managers in the US Department of Energy Office of Energy Efficiency and Renewable Energy have been using the Logic Model process since 1993 to help to communicate the progress and value of their programs to Congress, partners, customers, and other stakeholders (McLaughlin and Jordan, 1999). Also, the advanced technology program (ATP), a science and technology program under the control of the National Institute of Standards and Technology (NIST) that aims to enhance Americans' quality of life and their wealth through science and technology, applied the logic model to evaluate its effects (Ruegg and Feller, 2003). The aforementioned logic model was herein used as a basic conceptual model to establish a performance monitoring and evaluation management system for public R&D programs.

Rpmems

Concept of R&D Performance and its Monitoring

Research councils, which monitor and evaluate R&D performance of GRIs, need to clarify the concept of R&D performance and its monitoring before establishing the management system. The concept of R&D performance is defined to include not only research outputs, but also direct and indirect social and economic effects, such as converting such outputs into businesses, inputting them into other research projects and tasks, and technology transfer. This definition also includes records, information on outcomes, information on technologies, and statistics on performance as set out in the MOST's basic plan on the management and utilization of R&D performance.

Meanwhile, performance monitoring is defined as activities designed to conduct a long-term evaluation of outputs utilization status, and their technological, economical, and social utility, and thus to study, arrange, convert into statistics and analyze the value of R&D in an effort to manage R&D performance. With this definition, the concept of performance monitoring and evaluation needs to consider two aspects. Likewise, the concept involves monitoring and management as to whether or not GRIs monitor, investigate, arrange and analyze their own performance, and the performance evaluation of evaluator institutes that conduct absolute and relative evaluation of the qualitative and quantitative sizes of performance.

Considerations for Structuring RPMEMS

This research aims to set the directions of structuring the RPMEMS, and then to establish relations among the system's structure, scope, and factors. First, the performance of the RPMEMS should be considered. An effective and transparent RPMEMS should be structured to bolster the domestic industrial technology innovation capabilities through the maximization of the utilization of industrial technology R&D performance; the system should also be structured in a way that it encourages the R&D activities of GRIs and provides incentives for superior performance, thereby receiving the support of evaluatee institutes. Second, the system should also consider a linkage with the performance of upper-level policies. Likewise, it should be linked to the Act on Performance Evaluation for National R&D Programs, and so on and Performance Management, and to the policy objectives of NSTC. Third, the system should consider the evaluation institutes' own performance and its relations with the existing evaluation system. The mentioned act (on Performance Evaluation and Management for National Research & Development (R&D) Programs' in 2005) provides that the central administrative agencies and research councils, as deemed necessary, may establish, or designate a specialist institute (an institute specializing in research management). Thus, a system needs to be structured so as to substantiate the evaluation institutes' work and bolster their profile through their own specialist institutes.

Structure and Role of RPMEMS

The current entire process of national R&D from the start to the finish is divided into the R&D system of GRIs and the evaluation system for research councils. The RPMEMS aimed at increasing the utilization of R&D performance is positioned in the last stage. The three systems relating to R&D are shown in Fig. 2. Figure 2 shows a logic model-based structure involving operators of these systems, and their activity objectives and details.

The RPMEMS is different from the existing R&D evaluation system in that it involves users of outcomes in the group of participants, thereby emphasizing the utilization aspect of research results as mid- and long-term outcomes. The system monitors the utilization of these outcomes, inputs this information into the evaluation and management system, and enhances the utilization of outcomes by considering users of outcomes in

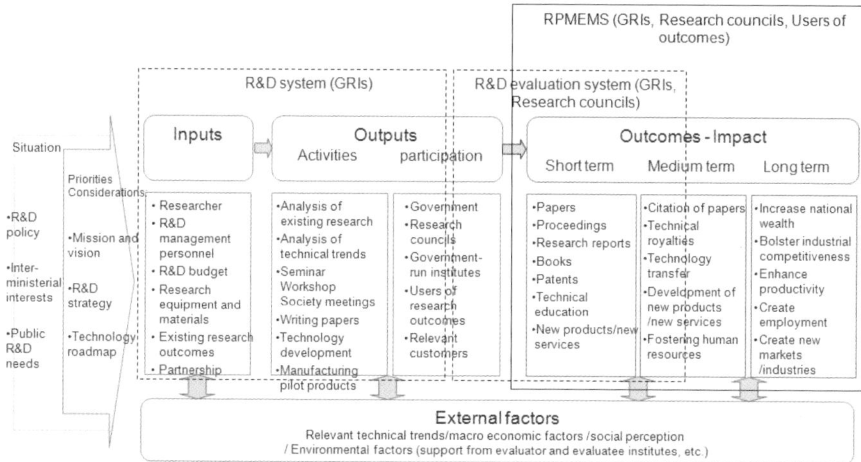

Figure 2: Targets and scope of RPMEMS.

advance from GRIs' R&D stage; if necessary, the system considers institutional support for utilizing outcomes and monitoring such utilization.

Notably, the RPMEMS is divided into three categories. First, as shown in Fig. 3, major operators of the system, as organizations relating to monitoring and evaluation, include upper evaluator institutes (NSTC/MOST), research councils, relevant GRIs, and networks of using outcomes (corporations, GRIs, public institutes, universities, general public, and users of outcomes).

Second, these operators interact with one other as they play their own roles. The uppermost evaluator institutes plan comprehensive evaluation, monitor outcomes, and establish evaluation-related major regulations and implementation procedures, as well as delegate their authority to lower evaluator institutes and supervise them, plan strategic technology development for evaluatee institutes on the basis of evaluation results, allocate budgets, and establish support policies. Evaluator institutes establish the monitoring, evaluation, and management system, to manage and evaluate their performance, and

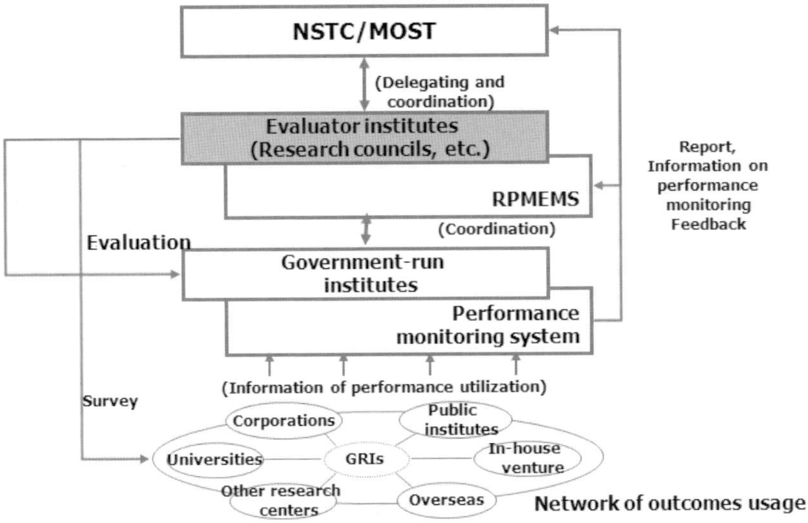

Figure 3: Factors of RPMEMS and their relationships.

report to upper evaluation institutes, as well as adjust the evaluation system taking into account the evaluatee institutes' difficulties. Also, GRIs — the key players in the system — establishes the performance monitoring system for utilizing outcomes, managing their performance, providing feedbacks, and receiving evaluation. Notably, they establish official/unofficial networks for using outcomes aimed at an enhanced use of outcomes, managing their performance and gathering the information of using outcomes. Lastly, the network comprising users of outcomes provides GRIs with the information of performance monitoring, namely, the direct and indirect effects of actually utilized outcomes.

The monitoring and evaluation process, the last factor of the RPMEMS, provides operators of the system with basic procedures for extracting information of outcomes and evaluation activities as guidelines on specific evaluation systems, details, methods, procedures, and time with a view to monitoring and evaluation. The next section presents a detailed review of this process.

Performance Monitoring and Evaluation Process

The performance monitoring and evaluation process consists of evaluation institutes, execution institutes' action process, extraction process designed to extract performance monitoring information (the target to be evaluated), and guidelines on execution.

The guideline on performance monitoring and evaluation includes the objects of monitoring, period, contents (evaluation items, methods, and other extraction process), and evaluation system and process (action process). The guideline on performance monitoring and evaluation is briefed as follows.

- Division of outcomes: short-term, intermediate, and long-term outcomes
- Target: R&D projects
- Period: every year for 5 years after completing a project (the evaluated institute will choose 3, 4, and 5 years according to the nature of the project)
- Item: scope of the defined R&D outcomes
- Evaluators: GRIs evaluate themselves (reporting evaluation results), or evaluator institutes evaluate reports by GRIs on utilization (monitoring) of outcomes.

The action process is an execution procedure by which the interaction for evaluation between evaluator institutes and evaluatee institutes is established by time; currently, each evaluation institute has an action process by which to evaluate GRIs, enabling it to conduct both the action process for performance monitoring and evaluation, and the existing process simultaneously, or to conduct the process differently. However, since the Performance Monitoring, Evaluation, and Management System is a new system to be introduced, evaluator institutes are advised to evaluate evaluatee institutors according to the existing evaluation process, thereby easing evaluatee institutes' workload. Figure 4 shows an example of the action process of the performance monitoring and evaluation.

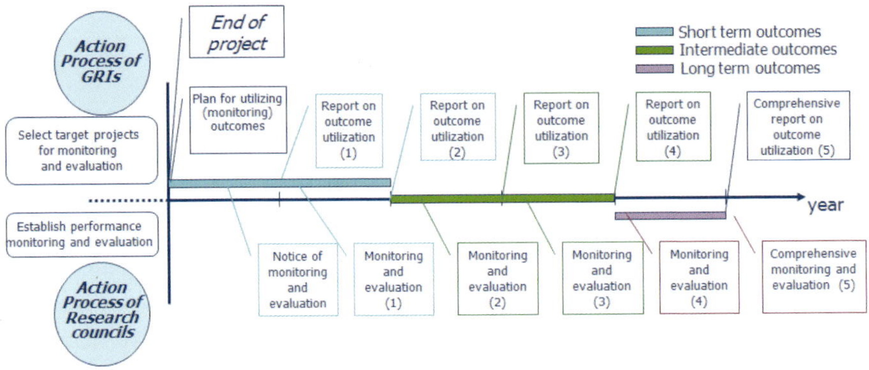

Figure 4: Example of action process for performance monitoring, evaluation, and management.

Lastly, the extraction process for performance monitoring information is needed for the evaluatee institutes — GRIs — to determine how and how much their research outcomes are utilized by users. Since the outcomes and their utilization may differ according to the characteristics of evaluatee institutes, flexibility should be exercised; however, a standardized system is needed in order to structure databases on the information of using outcomes at the public levels, and to utilize outcomes and conduct a relative evaluation of evaluate institutes.

Based on the definition of R&D performance monitoring, the information is divided into the information for management, and the information for performance monitoring and evaluation. The former means an infrastructure to enable to each evaluatee institute to monitor, arrange, and analyze quantitative and qualitative data concerning the utilization and spread of research outcomes. The latter means the information on research outcomes and on their utilization; it is divided into the performance category including record outcomes, research outcomes, information on outcomes, and technical information, and into the utilization category including the scientific and technical utilization of outcomes (technical support, etc.), and economic utilization of such outcomes (technology transfer, etc.) in an effort to establish a process for extracting the information on monitoring.

Methods for Performance Comparison in RPMEMS

Previous section presented the RPMEMS model, but RPMEMS alone cannot make it possible to conduct relative comparisons among GRIs. Given the nature of monitoring and evaluation, the following should be considered to conduct relative performance measurement, First, factors for monitoring performance are very diverse; thus in order to evaluate performance using objective figures, weights, or priorities of items should be determined for aggregation. Second, each item has different units such as number of cases, dollars, number of citations, and so on. Thus, it should be considered how to aggregate items with different units. Third, the periods for projects being monitored are different from each other. For projects performed in the same periods, the periods need not be considered, but for projects performed in different periods, the periods should be considered in order to ensure a fair evaluation. However, the existing method for relative evaluation on GRIs which consists of the management performance of 30% as a qualitative evaluation and the research performance of 70% as a quantitative evaluation cannot accommodate the above-mentioned consideration. This section presents two alternatives for performance comparison in the RPMEMS model.

Self-Weighting: Current Practice

The existing method called self-weighting is a method by which GRIs autonomously determine weight of items in each section, and have external experts (using the existing institute evaluation group) conduct qualitative and quantitative evaluation. In this way, the existing system applied in the evaluation of institutes can be utilized, thereby improving the administrative and procedural convenience and the evaluated institutes' adaptation to the system. Also, the existing institute evaluation group members can be used, offering the advantage of reducing evaluation time and costs. However, this method can be used only if the number of initial target projects being monitored is small, but may hurt the validity and rationality of evaluating the projects and GRIs. Also, if the number of target projects rapidly increases, this will take a large number of evaluators, increasing their evaluation workload rapidly, and making it difficult to ensure a strict and fair evaluation. This will also require a vast quantity of administrative efforts on the part of the research council. Thus, this method may be applied at the initial stage of monitoring and evaluation with its application time set differently from the institute evaluation.

The greatest weakness of this method is that, since the evaluation items and weights by item are determined by GRIs, it is impossible to conduct a relative evaluation of the institutes involved. Thus, this research presents the following two methods as alternatives.

AHP-Based Evaluation

Evaluation method and framework

The first alternative is to employ the AHP to determine the weights of factors for performance evaluation. The AHP proposed by Saaty (1980) is a technique designed to resolve problems involving MCDM with multiple evaluation criteria for multiple alternatives. It has also been effectively used in the evaluation of R&D projects. The AHP consists of five steps: (1) hierarchical model construction, (2) pair-wise comparisons, (3) local priority vectors, (4) consistency check, and (5) final priorities. For more information on the AHP, see the text by Saaty (1980).

Although the AHP has widely been used as a method for preevaluation (selection) problem in the field of R&D evaluation, recent years have seen an increase in its use for postevaluation (performance measurement). The basic principle is to derive priority weights of evaluation factors and apply them equally to all GRIs. The values of evaluation factors for each GRI are the aggregated sum of the corresponding factors of the projects under the control of the GRI.

The AHP-based evaluation has the following characteristics in terms of the above-mentioned three considerations. First, a single measure, priority weights of GRIs is derived to aggregate various evaluation factors. Second, variables with different units can be normalized based on the utility index. Third, an annual average concept is employed for projects with different monitoring periods. Figure 5 shows the AHP model for the relative evaluation of GRIs. The criteria consist of items of the highest-level classification category and the subcriteria represent items of the middle-level classification category. At the bottom, each GRI is included in the model as an alternative.

Example

An example is presented here to illustrate the AHP-based evaluation. First, pairwise comparisons have been made based on the

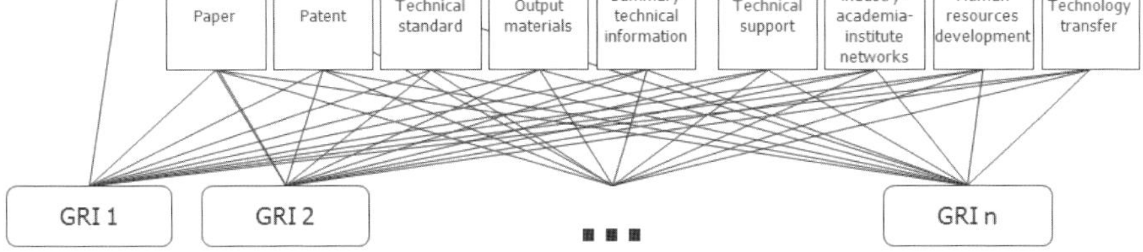

Figure 5: AHP model for relative evaluation of GRIs.

judgment of experts. Table 1 shows the pairwise comparison matrix among criteria with respect to the goal as an example. In this way, pairwise comparisons were conducted among subcriteria with respect to "Performance" and "Utilization". The derived final priority weights are shown in Table 2.

Pairwise comparisons among GRIs do not have to be done since the data for subcriteria are collected from the RPMES. Since the values of evaluation factors have different units, however, they should be normalized. For this purpose, the utility index is employed here. It is defined as the following:

$$u(X) = \frac{X - X^-}{X^+ - X^-}$$

where X^+ is the best value; X^- is the worst value; X is the real value. The obtained utility indices range from 0 to 1, 0 for minimum value and 1 for maximum value. Since the sum of the priority weights in the AHP equals to one, the utility indices are renormalized by dividing them by the sum.

The data for subcriteria of hypothetical 10 GRIs were randomly generated. Table 3 shows the obtained priority weights of GRIs for each factor.

The final priority weights of GRIs can be computed by weighted sum of priority weights of factors (Table 2) and priority weights of GRIs for each factor (Table 3). Table 4 shows the final evaluation scores and ranking of GRIs.

Table 1: Pairwise comparison matrix among criteria.

	Infrastructure	Performance	Utilization	Priority Weight
Infrastructure	1	3	1	0.4286
Performance		1	1/3	0.1428
Utilization			1	0.4286

Table 2: Priority weights of evaluation factors.

First Level	Local Priority Weight	Second Level	Local Priority Weight	Global Priority Weight
Infrastructure	0.4286			0.4286
Performance	0.1428	Paper	0.1779	0.0253
		Patent	0.3491	0.0497
		Technical standard	0.3900	0.0555
		Output materials	0.0415	0.0059
		Summary of technical information	0.0415	0.0059
Utilization	0.4286	Technical support	0.1021	0.0438
		Industry-academia-institute networks	0.3479	0.1491
		Human resource development	0.1021	0.0438
		Technology transfer	0.4479	0.1920

Table 3: Obtained priority weights of GRIs for each factor.

	Infra.	Paper	Patent	Technical Stand.	Output	Summary	Technical Sup.	IAI Network	Human	Technology
GRI 1	0.0350	0.1178	0.0531	0.0616	0.0217	0.0006	0.0693	0.0844	0.0648	0.0658
GRI 2	0.1359	0.0932	0.0835	0.1417	0.0645	0.0981	0.0341	0.0809	0.1349	0.0405
GRI 3	0.0332	0.0454	0.1574	0.1965	0.1935	0.0523	0.0362	0.0554	0.1308	0.0934
GRI 4	0.0147	0.1829	0.2119	0.0039	0.0518	0.1690	0.1438	0.0626	0.1431	0.0697
GRI 5	0.1863	0.0588	0.0274	0.1833	0.1423	0.1150	0.1096	0.0319	0.1440	0.1034
GRI 6	0.0775	0.1022	0.0143	0.1511	0.0690	0.1334	0.1939	0.0363	0.1032	0.2326
GRI 7	0.1775	0.0113	0.1846	0.0540	0.2095	0.0800	0.2562	0.1972	0.0193	0.1058
GRI 8	0.1483	0.0089	0.0913	0.1102	0.1069	0.0897	0.0359	0.1356	0.1102	0.1268
GRI 9	0.0493	0.2048	0.0925	0.0888	0.0009	0.1627	0.0641	0.1817	0.1086	0.0205
GRI 10	0.1423	0.1748	0.0838	0.0090	0.1399	0.0991	0.0569	0.1340	0.0412	0.1413

Table 4: Final evaluation scores of GRIs.

GRI	Final priority Weight	Rank
GRI 1	0.0553	10
GRI 2	0.1008	6
GRI 3	0.0691	8
GRI 4	0.0583	9
GRI 5	0.1301	2
GRI 6	0.1092	5
GRI 7	0.1520	1
GRI 8	0.1265	3
GRI 9	0.0754	7
GRI 10	0.1229	4

DEA-Based Evaluation

Evaluation method and framework

DEA is a nonparametric productivity measurement technique which does not assume the relative importance of input factors and output factors, and a production function. DEA has the following advantages in connection with R&D performance evaluation (Wang and Huang, 2007). First, it is not easy to determine the weights of the input/output factors with regard to R&D; however, DEA automatically determines the weights that maximize each DMU's efficiency, thereby eliminating the need to predetermine the weights. Such a flexible and endogenous weighting system of DEA is called the "benefit of the doubt" approach (Cherchye *et al.* 2007). Second, given the nature of R&D, the relationship between input and output is difficult to express in terms of a function; however, DEA is a nonparametric method that does not have a particular function. Third, R&D involves diverse input and output factors; DEA can reflect them without limiting their numbers. Such advantages of DEA allow it to be widely used in the evaluation of the performance at diverse levels of R&D such as nations (Lee and Park, 2005; Kocher *et al.*, 2006; Wang and Huang, 2007), program (Lee *et al.*, 2008b, university (Korhonen *et al.*, 2001; Feng *et al.*, 2004; Cherchye and Vanden-Abeele, 2005), project (Korhonen *et al.*, 2001; Garg *et al.*, 2005; Eliat *et al.*, 2006), and task team (Paradi *et al.*, 2002).

This study also presents DEA model for measuring performance of GRIs. In connection with the aforementioned considerations in conducting relative evaluations of GRIs, DEA offers the following specific advantages. First, since DEA considers the monitoring periods as an input factor, yearly variation can be compensated. Second, the efficiency score resulting from DEA is unit-invariant; thus, DEA can consider diverse variables with different units. Third, DEA can automatically assign high weights to those items in which each project has strengths, enabling the given performance contents to be utilized to the greatest advantage. This feature can reflect the aspect of GRIs giving a high weight to their favored items in the case of self-weighting.

Figure 6 shows the DEA-based relative evaluation framework. Considered as input factors are R&D expenses, the number of researchers, and monitoring period. The R&D expenses and the number of researchers were selected as proxy for capital and labor, respectively, and these are typical input factors in similar research. Monitoring period was selected in order to mirror different periods by project in the evaluation. As shown in Fig. 5, the period leading to a time of relative evaluation is considered as an input factor, thereby eliminating the influence of yearly variation.

Every project under every GRIs is evaluated together by DEA. Then, the performance of a GRI is measured based on those of its projects being monitored. However, simple comparison based on average efficiency scores does not have

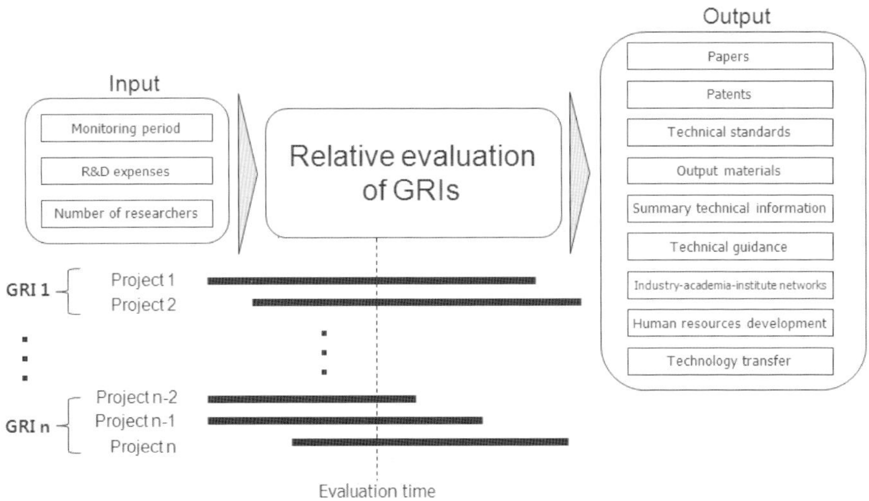

Figure 6: DEA framework for relative evaluation of GRIs.

statistical validity since theoretical distribution of the efficiency scores in DEA is unknown (Lee *et al*., 2009). Thus, Kruskal–Wallis test, which is a nonparametric method for comparing means, is run to compare performance of GRIs.

Example

To show how the DEA-based evaluation is conducted, an example is given on ten hypothetical projects of three GRIs. For the convenience of illustration, only monitoring period and R&D expenses are considered as input factors. Also, output factors include patents and papers. The basic DEA model, input-oriented CCR model, is employed for measuring efficiency of projects. Table 5 shows the data for DEA, and the results are shown in Table 6.

Then, Kruskal–Wallis test was conducted to compare performance of GRIs. Table 7 shows the results. It is shown that GRI 1 has the best performance, followed by GRI 2 and GRI 3.

Table 5: DEA data.

		Input		Output	
GRI	Project	Monitoring Period	R&D Expense	Papers	Patents
GRI 1	Project 1	4	1,400	30	8
	Project 2	4	1,520	41	4
	Project 3	3	1,800	30	5
	Project 4	3	1,800	53	5
GRI 2	Project 5	5	1,590	38	1
	Project 6	5	1,730	37	3
	Project 7	4	1,370	30	2
GRI 3	Project 8	5	1,800	30	5
	Project 9	4	1,270	23	3
	Project 10	3	1,040	16	1

Table 6: DEA results.

GRI	Project	Efficiency Scores	Rank
GRI 1	Project 1	1.00	1
	Project 2	0.92	3
	Project 3	0.89	4
	Project 4	1.00	1
GRI 2	Project 5	0.81	5
	Project 6	0.73	7
	Project 7	0.74	6
GRI 3	Project 8	0.65	9
	Project 9	0.66	8
	Project 10	0.52	10

Table 7: Results of Kruskal–Wallis test.

	GRI 1	GRI 2	GRI 3
Mean rank	8.50	5.00	2.00
P-value		0.018	
Comparison		GRI 1 > GRI 2 > GRI 3	

Discussions

The proposed two alternatives, DEA- and AHP-based evaluation, can overcome the limitations in the current practice, but it is hard to make judgment about which of the two is better due to the existence of their pros and cons. The decision on which method to employ totally depends on the context of monitoring and evaluation. The context variables include the need for fixed weights, the need for expert judgment, the need for full-ranking, and the number of projects being monitored and evaluated, information loss, and the structure of evaluation factors. First, if there exists accepted views on preference between input and output variables, the AHP is preferred since it fixes weights of variables included in the hierarchy model. However, when it is not easy to determine the weights of the input/output factors or fixing weights can cause opposition of GRIs, DEA is better since it automatically determines the weights that maximize efficiency of projects under GRIs. With respect to the first issue, second, expert judgment is required in the AHP to determine fixed weights, but it does not required in DEA. Third, using the AHP is appropriate when full-ranking is required. Since all efficient DMUs who have 100% efficiency are considered equally good in DEA, DEA cannot perform full ranking without modification of the DEA models (Ader *et al.*, 2002). When the number of GRIs is small, fourth, the AHP performs better since the number of pairwise comparisons that need to be made in the AHP dramatically increases when the number of alternatives increases. On the other hand, when the number of DMUs is small in DEA, a large portion of the DMUs will be identified as efficient so the efficiency discrimination is lost (Cooper *et al.*, 2000). Thus, DEA should be used when the number of projects being monitored is relatively large. The rule of thumb is that the number of DMUs (projects) should be at least three times larger than the combined number of inputs and outputs (Banker *et al.*, 1989) or more than the product of the numbers of inputs and outputs (Boussofiane *et al.*, 1991). Fifth, information loss occurs in using the AHP since the data at the level of projects are aggregated into those at the GRI level. On the other hand, DEA does not exhibit information loss since DMUs are projects, not GRIs. Finally, if the structure of evaluation factors is composed of multiple levels, DEA cannot accommodate it while the AHP models it as a hierarchy. Table 8 summarizes the guidelines mentioned above on selection between the two alternatives.

Conclusions

This study proposed the model of RPMEMS for effectively utilizing and spreading national R&D outcomes. The model for RPMEMS based on the logic model was presented with its scope, structure, role, and process. Also, the two methods for relative

Table 8: Guidelines on selection of methods for relative comparison.

	DEA	AHP
Weights	Flexible	Fixed
Expert involvement	Not required	Required
Full-ranking	Not required	Required
Number of projects	Large	Small
Information loss	Not occur	Occurs
Structure of evaluation factors	Single level	Multi-level

comparison, DEA and AHP, were proposed, and their pros and cons were discussed.

The contribution of this study is twofold. First, the proposed RPMEMS model can be practically implemented and operated for performance monitoring, evaluation, and management of national R&D. Although it is targeted to the Korean research councils, it can be used for other institutions in charge of national R&D in another country by customizing it according to their own context; or it may be employed a reference model. Second, the DEA- and AHP-based frameworks enable relative comparisons to be made in conjunction with performance monitoring and management. They overcome the limitations of the current practice and can be considered as promising alternatives.

Nonetheless, this study is subject to some limitations that will serve as fruitful avenues for future research. The RPMEMS model and evaluation methods should be modified in a way that the model and methods can be accepted by evaluated institutes, GRIs. Since the model and methods have been developed from the perspective of evaluator institutions, they may not be able to accommodate different views and conditions of GRIs. Thus, this study needs to be further developed by reflecting such diverse views and conditions. Also, the refinement of the model requires it to be implemented and tested in practice.

References

Ader, N., Friedman, L., and Sinuany-Stern, Z. (2002). Review of ranking methods in the data envelopment analysis context. *European Journal of Operational Research*, **140**: 249–265.

Banker, R. D., Charnes, A., Cooper, W. W., Swarts, J., and Thomas, D. A. (1989). An introduction to data envelopment analysis with some of its models and their uses. *Research in Governmental and Nonprofit Accounting*, **5**: 125–163.

Bickman, L. (1987). The functions of program theory. In L. Bickman (Ed.), *Using Program Theory in Evaluation, New Directions for Program Evaluation*. San Francisco: Jossey-Bass.

Boussofiane, A., Dyson, R. G., and Thanassoulis, E. (1991). Applied data envelopment analysis. *European Journal of Operational Research*, **52**(1), 1–15.

Cherchye, L., and Vanden-Abeele, P. (2005). On research efficiency: A micro analysis of Dutch university research in economic and business management. *Research Policy*, **34**: 495–516.

Cherchye, L., Moesen, W., Rogger, N., and van Puyenbroeck, T. (2007). An introduction to "benefit of the doubt" composite indicators. *Social Indicators Research*, **82**(1): 111–145.

Cooper, W. W., Seiford, L. M., and Tone, K. (2000). *Data Envelopment Analysis: Theory, Methodology, and Applications, References and DEA-Solver Software*. Boston: Kluwer Academic.

Eliat, H., Golany, B., and Shtub, A. (2006). Constructing and evaluating balanced portfolios of R&D projects with interactions: A DEA based methodology. *European Journal of Operational Research*, **172**(3): 1018–1039.

Feng, Y. J., Lu, H., and Bi, K. (2004). An AHP/DEA method for measurement of the efficiency of R&D management activities in universities. *International Transactions in Operational Research*, **11**: 181–191.

Garg, K. C., Gupta, B. M., Jamal, T., Roy, S., and Kumar, S. (2005). Assessment of impact of AICTE funding on R&D and educational development. *Scientometrics*, **65**(2): 151–160.

Graves, S. B., & Langowitz, N. S. (1996). R&D productivity: A global multi-industry comparison. *Technology Forecasting and Social Change*, **53**(2), 125–137.

Kellogg Foundation. (2004). *Logic Model Development Guide*.

Kocher, M. G., Luptacik, M., and Sutter, M. (2006). Measuring productivity of research in economics: A cross-country study using DEA. *Socio-Economic Planning Sciences*, **40**(4): 314–332.

Korhonen, P., Tainio, R., and Wallenius, J. (2001). Value efficiency analysis of academic research. *European Journal of Operational Research*, **130**: 121–132.

Lee, H., and Park, Y. (2005). An international comparison of R&D efficiency: DEA approach. *Asian Journal of Technology Innovation*, **13**(2): 207–221.

Lee, H., Lee, C., Seol, H., and Park, Y. (2008a). On the R&D Priority Setting in Technology Foresight: A DEA and ANP approach. *International Journal of Innovation and Technology Management*, **5**(2): 201–220.

Lee, H., Park, Y., and Choi, H. (2008b). Comparative evaluation of performance of national R&D programs with heterogeneous objectives. *European Journal of Operational Research*, DOI:10.1016/j.ejor.2008.06.016

Lee, M., Son, B., and Om, K. (1996). Evaluation of national R&D projects in Korea. *Research Policy*, **25**(5): 805–818.

McLaughlin, J. A., and Jordan, G. B. (1999). Logic models: A tool for telling your program's performance story. *Evaluation and Program Planning*, **22**(1):, 65–72.

Nakamura, Y., and Tsuji, M. (2004). Study on the allocation of R&D investment: Method of evaluating and selecting R&D projects for investment. *International Journal of Innovation and Technology Management*, **1**(1): 55–74.

Paradi, J. C., Smith, S., and Schaffnit-Chatterjee, C. (2002). Knowledge worker performance analysis using DEA: An application to engineering design teams at Bell Canada. *IEEE Transactions on Engineering Management*, **49**(2): 161–172.

Ruegg, R., and Feller, I. (2003). *A Toolkit for Evaluating Public R&D Investment Models, Methods, and Findings from ATP's First Decade*. Advanced Technology Program.

Saaty, T. (1980). *The Analytic Hierarchy Process*. New York: McGraw-Hill.

Wang, E. C., and Huang, W. (2007). Relative efficiency of R&D activities: A cross-country study accounting for environmental factors in the DEA approach. *Research Policy*, **36**(2): 260–273.

Wholey, J. S. (1983). *Evaluation and Effective Public Management*. Boston: Little, Brown.

Wholey, J. S. (1987). Evaluability assessment: Developing program theory. In L. Bickman (Ed.), *Using Program Theory in Evaluation, New Directions for Program Evaluation*. San Francisco: Jossey-Bass.

Chapter 10

New Product Development: Balancing between R&D And Marketing

*Ahmet Ilker Soydan** *& M. Atilla Oner*†

Introduction

Trade-off between marketing and research-development has always been a dilemma for most managers. Balancing the budgets between different departments and estimating the future profits and customer base as a result of this action has remained a "challenging task" throughout the contemporary business history.

Managers and researchers have come up with many tools to enhance their understanding of the business/market structure and to ease the strategy formulation process (Scherer, 1967; Souder, 1972; Baker and Freeland, 1975; Zoltners and Sinha, 1980; Lee, 1985; Gill, 1992; Rink *et al.*, 1999; Shapiro, 2003; Jenkins, 2005). However, these tools have fallen short in some periods due to difficulties in their applicability.

The importance of standard strategic management tools and their inadequacy have been stressed by Graham and Ariza (2003) not only because of communication and execution but also fundamental limitations of methodology. Complexity often determines the tools to be used. In a highly complex demand estimation projects, system dynamics (SD) proves quite

*SIEMENS A.S. Turkey, Corporate Development Strategies.
†Yeditepe University, Management Application and Research Center.

useful to analyze the problem. In the model structure of Graham and Ariza (2003) one sees the quite intensive and complex interaction of marketing and R&D.

The purpose of the paper is to investigate and find the sustainable allocation between marketing and R&D so that a firm can keep a sustainable level of customers and creating sufficient amount of revenue. While searching for allocation policies, special attention is paid to the timing of these decisions to synchronize the two activities. The paper attempts to contribute to the understanding and the management of a complex problem via policy design and analysis.

The chapter is structured as follows: the next section reviews the relevant literature to extract the model variables. Section "Methodology" describes the methods used. Section "Simulation Model" introduces the proposed causal loop diagrams and identified stock and flow variables. Section "Testing and Validation" describes different tests that are used to validate the model. Section "Simulations and Behavior Discussions" discusses simulations and resulting behaviors. "Policy Design and Evaluation" covers the tests, evaluation, and improvement of policies. The last section concludes the chapter.

Literature Review

In this section, we review literature on R&D resource allocation from the different perspectives in order to identify and decide on the components of our model.

The phenomenon of research and development rivalry in a dynamic profit maximization framework was analyzed by Scherer (1967) who predicted the market structural conditions most conducive to rapid technological progress by formulating models of the firm's optimal R&D resource allocation problem. The author listed factors affecting the revenues from successful R&D project completion as the **date of completion**, the **quality of the end product** (related to its ability to satisfy existing or latent demands), **obsolescence time**, and **the reactions of rivals.**

Factors affecting the costs of an R&D project were listed as the **state of technology**, the **quality of the end product**, and the **speed of the development**. Given the parameters assumed in the model, 3 years was reported as the minimum time to product introduction (with infinitely high development costs), whereas 8.5 years was the development schedule which minimizes undiscounted total costs.

Developing products faster, better, and cheaper than competitors has become critical to success in many markets (Ford and Sterman, 1998). However, in order to do this, there should be adequate tools and mechanisms to forecast the demand for the product so that quality and awareness can be adjusted accordingly. In addition, to develop products faster, better, and cheaper may require huge initial investments in the underlying business processes. This process also depends on the sector that the company is operating in. Obviously, the characteristics of the sector define the need for innovations and marketing.

In a classification of high, medium, and low technologies (OECD, 1996), one would expect that marketing and R&D budgets for a high-tech industry firm (e.g., airplane, medical industry), for a low-tech industry firm (e.g., tobacco, leather) and for a mid-tech industry firm (e.g., chemical, plastics) would be different depending on their elasticities to price, demand, and supply. Trott (2002) gave a useful classification scheme of New Product Development (NPD) activities across different industries. New product quality level goals have been argued to differ depending on the technology level of the sector (Scherer, 1967).

One may also expect that R&D triggers the company growth due to increased demand in the market. Kameoka and Takayanagi (1999) demonstrated that R&D became one of the most essential driving forces for corporate growth as well as survival in crisis periods. They claimed that R&D expenditures as input resources were gradually converted into technology stock with a certain time delay. Anderson and Reynolds (1997) pointed out that demand articulation was a process of converting the customer's vague wants into a set of research and development projects. This theory, they said, is completely nonlinear, market-driven, and involves foreseeable targets as Dickinson (1999) stressed manager's obligation to allocate resources to diverse financial instruments in the proper proportions to meet the customer's needs. One has to balance risk and reward for this portfolio management.

Company managers and scholars have increasingly come to realize that design and engineering play an important role in delivering products to the final customer (Repenning, 2000). Hauser (1997) mentioned that R&D provides the science and technology, which firms use to serve tomorrow's customers profitably. Many managers, consultants and researchers have argued that, to succeed in the 21st century, R&D should be market-driven. The increasing integration of marketing and R&D has been shown to bring increasing success, especially on new product performances (Leenders and Wierenga, 2002). This should be done after acquiring required level of resources so that building the integration would make sense and effect on business value.

Confessore and Mancuso (2002) discussed that in a dynamic R&D decision process, the impact produced by a firm's R&D expenditure at a given period, produces both positive and negative impacts on the stream of firm's success. The total R&D expenditures and government subsidies including basic research in universities build up the national R&D competitiveness. It becomes crucial when these budgets are allocated in the companies, which constitute the microeconomy. Investigating the link between national competitiveness and R&D, Papadakis (1995) found out that R&D intensive industries have high competitiveness when compared to the others and that the level of R&D investment has an impact on economic/competitive performance. She also stressed the fact that technology push forces radical innovations in new emerging markets and demand pull generates an emphasis on process innovation in markets with proven demand. It is of benefit to notice that there is a virtuous cycle where more investment in R&D makes nations, brands, and companies more competitive in the next time around.

Resource allocation decisions between exploration (research) and exploitation (development) were examined by Garcia *et al.*, (2003) with the help of a SD model by focusing on four different factors: **resource availability, exogenous competition, aging of knowledge bases,** and **adaptive capacity.** They made three normative suggestions for new product development management in the proportional choice decision-making for explorative and exploitative activities. They also pointed out to the fact that a healthy institution always should allocate some slack resource into the exploration activities even in the turbulent periods. This is due to the erosion of the knowledge stock with time and eventually one needs to feed in new investments.

The endogenous structure for the R&D procyclical behavior in G7 countries was investigated by Wälde and Woitek (2004). They tested the prediction that resource allocation to R&D should be negatively correlated to economic growth in terms of Gross Domestic Product (GDP). They found that some of the R&D activities were accelerated during the economy's downturns during which marketing budgets are pruned. In the present study, the opposite of this was tested.

Methodology

Applicability of SD to the Problem

"The desire to know the future is a deep human psychic need" (Sherden, 1998; cited in Lyneis, 2000). In order to satisfy this psychic need, managers use(d) many tools. SD is one of the most powerful tools for forecasting in complex structures.

One of the major advantages of SD over the other methodologies is its ability to help the modeler understand the underlying structure of the current problem at hand. The reason that

SD becomes superior rather than the other methods is that it deals with causalities and nonlinearities which are neglected, ignored, or even feared by most of the static analysis. By applying SD, one can enhance the usefulness of the model to address and analyze problems to provide more significant, rational, and pertinent policy recommendations. The real value of an SD model should be to eliminate problems by changing the underlying structure of the system rather than anticipating and reacting to the environment. A SD model, this way, interacts with the environment and gives/alerts feedback for structure changes (Sterman, 2000). A SD Model provides more reliable forecasts than the statistical (non-structural) models and tries to understand the underlying structure that create the data stream (Lyneis, 2000).

The problem discussed in the present chapter is a dynamic problem because the trade-off between the marketing and R&D fractions may change each and every year. This leads to different sales figures, which would lead to different profitability figures. The firms have to have new products to attract customers as well as marketing to support these products and the old ones. However, the budget of a firm is not limitless. Therefore they need to have a policy mix for these issues. Besides, 1 year's solution may not be relevant for the next year.

Problem Definition, Scope, Time Horizon, and Purpose

Companies may suffer from excess resource allocation, such as over-investment in the market expenditures though people do not have sufficient purchasing power and over investment in R&D though the market is satisfied with the new products. The compromise and harmony between the marketing and R&D determines the success of the firm in terms of customers, which would bring revenue.

The electro market demand fluctuates year by year due to GDP growth, which brings severe planning and strategy formulation difficulty in terms of marketing and R&D. Therefore, the fractions to be shared by R&D and marketing changes every planning year for the companies. Depending on the changes in the effectiveness of resource use, one can say that over- and under-allocation is a widely faced situation.

Maintaining the sustainable success which is mainly ***keeping the customer base at a sustainable level*** and the market share for the electro-business can be handled via the marketing or R&D activities such as quality improvement or new product launches. GDP growth can also affect the market and decisions drastically. However, it is not aimed to model GDP endogenously in the present work.

The general framework of the model includes decisions such as marketing and R&D (for both the average quality of the products and aging new product launches). Production, inventory, backlog, and issues such as delivery delay, competitors' reactions, and licensing are not considered.

On the other hand, product attractiveness, quality, and awareness are other effects on this success. One can see that there are many trade-offs that a CEO should consider and manage differing levels of stocks of the firm. This brings the problem of understanding and controlling complexity in the market structure.

Considering the delay of R&D expenses on R&D know-how, it can be noticed that it takes 5 or so years to form or add to such a stock. Hence, we decided to use a time horizon of 20 years to study the impact of different product generations.

The purpose of the model is to investigate and find the sustainable allocation trends in marketing and R&D so that a firm can keep at a sustainable level of customers. Customers are assumed to be the main resource on which the firm builds all of the success indicators. While looking for these allocation trends, special attention is paid to the timing of these allocation decisions, namely the harmony between these two departments. One can see the model boundary chart in Fig. 1.

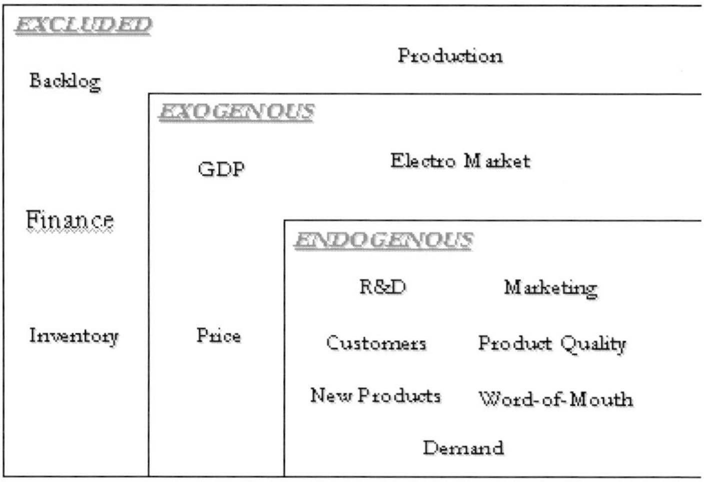

Figure 1: Model boundary chart.

The model comprises of mainly the following issues:

1. Customer Accounting Management
2. Product Portfolio Management
3. New Product Development
4. R&D Quality and Manufacturing

One can find relevant sectors and variables in the model related with these topics. In the next sections, the dynamic formulation of the hypotheses in both causal diagram and stock-flow representations will be explained. Following this, validation tests are done and then the policy design will be clarified. Finally, we conclude with results and further research.

Formulation of Dynamic Hypotheses

The main aim of the model is to provide and explanation for the allocation trends in marketing and R&D departments of a high-tech firm to achieve a long-term success in customer base and revenue. We target to explain and investigate the reasons behind these investment patterns in details. In particular, lack of harmony and timing of the investments in these functions are believed to create the problems concerned in this project which is the departure point for the dynamic hypotheses.

One way to evaluate these hypotheses is through a look at the timings of the allocation and its effect on the relevant stock. Marketing may have a short-term effect whereas R&D has a longer term effect. Especially in terms of R&D, it takes specific longer times to create an R&D know-how and functionality in the department. One should not expect to get a quick response from R&D which is ignored by most of the business environments. Time delays should be taken into consideration.

Simulation Model

The model in this chapter is intended to represent the development of resource (budget) allocation of a high-tech company. The goal of the chapter is to be a guide for a manager of the firm to constitute healthy investment actions for marketing and R&D. This would be via changing and controlling certain values of certain parameters in the system. When we change these values, we should see an effect on the objective targets of the model. After observing the effects of decisions taken to different types of allocation trends, the user would help improve his or her understanding of the situation and structure that exists on his or her mental model.

The modelling process starts with determining some relevant key variables in the model to be explored for the problem at hand. The previous sections give an insight about these variables. The model boundary chart is developed to identify what is exogenous, endogenous, and excluded to determine the scope and boundary of the model. One can see a rough sketch of the sector relations reflecting the model structure in Fig. 2.

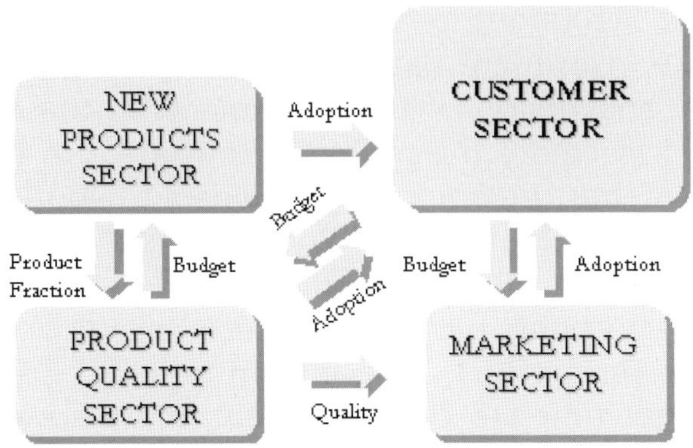

Figure 2: Bird-eye view of the model in terms of sectors.

Customer Sector works like a big engine for the model and provides Revenue for the survival and development of the other sectors. Product Quality Sector provides the Quality for the products and feeds this back to the Marketing sector for increasing **Awareness** which would bring more **Adoption of Customers**. New Product Sector sustains the R&D Budget by the help of its feedback to the Product Quality Sector where the main **R&D Budget** is determined.

Given the dynamic hypotheses and these sub-system charts, the model should represent a situation in which allocation gets an equilibrium value, which is in fact the sustainable level for each department that ought to work in harmony. The underlying feedback and stock-flow structure would help in capturing the model (i.e., the main causalities and equations).

The model assumes a laboratory environment where the data for the model are fictive, but close the reality. The next step in this model should be applying this to a real case with the actual data. However, as a starting point, this chapter would build confidence and ability to control the variables so that the next step gets much easier to handle. After the stock-flow structure is introduced, some policy designs would be created and analyzed for different situations.

Causal Loop Diagrams

The Causal Loop Diagram gives the general structure based on causalities and feedback loops. The interpretation of these loops and their combined complexity can be recognized easily. It may help separate the loops to two groups, namely Reinforcing and Balancing Loops. The red arrows show the Balancing Loops and the green ones show the Reinforcing Loops. One can see the schema of causalities starting from Fig. 3 and in the following figures. The storytelling can be made both ways. In other words, if there is a (+) causality between two variables, it is possible to say that "X increases Y" if X increases and "X decreases Y" if X decreases. Each one of the loops are explained and described in the following pages due the sector analysis.

It is relevant to start explaining the story behind a resource because of the fact that a resource has a meaning when time is frozen, as it is at the beginning of the story. The selection of the

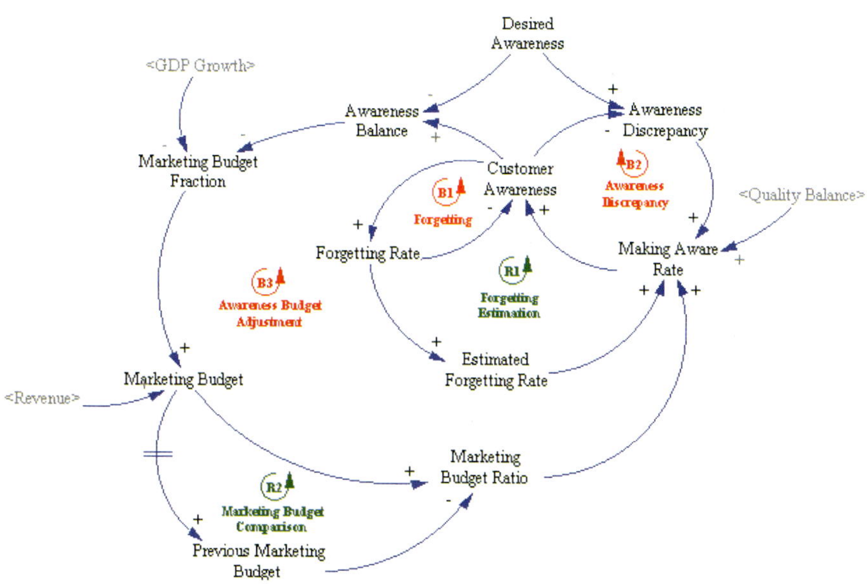

Figure 3: Causal loop diagram of the marketing sector.

stocks would be easier for the reader to understand in the following sections.

There are 4 sectors, namely Marketing, R&D Quality, R&D New Products, and Customers. These contain the major variables of the proposed model.

Marketing Sector

It is assumed that the company has a **Customer Base** which forms the sales and revenue figures for the survival of the company. The formation of the **Revenue** and **Customers** will be explained in detail later. The more revenue we have, the more **Marketing Budget** we would get. As the **Marketing Budget** increases, **Marketing Budget Ratio** is also increased because the ratio is the comparison of the current budget with the previous quarter's budget. When the company realizes that the **Marketing Budget Ratio** is increased, there occurs more opportunity for improvement in the **Customer Awareness**. This rate of **Making Aware** is also affected by the **Quality Balance** which is another variable in another sector. As the **Quality Balance** is increased, **Making Aware** rate is increased too. When customers see that the quality of the product is high, they get aware of it more. The more aware the customers are, the more forgetting there occurs. This **Forgetting Rate** is estimated within a time period and added to the **Making Aware** rate, so that the managers replace the depreciated **Awareness** with the new amount.

There is also a **Desired Awareness** that the company seeks for. **Awareness Discrepancy** is the difference between the actual and desired value for the awareness. The more **Awareness Discrepancy,** the more addition to the **Making Aware** rate. In addition, the company can see its awareness by dividing **Customer Awareness** by the **Desired Awareness** which is called **Awareness Balance**. As the **Awareness Balance** gets bigger, the budget gets smaller each time. The reason is that a customer can be maximum 100% aware of the product or service. Therefore this balance ratio and also the **GDP Growth** feed back to the budgeting decisions. Finally we come up with the new budget for marketing where we have started.

R&D-Quality sector

In order to determine the **R&D Budget**, **Old Product Fraction**, and **GDP Growth** are used. As the fraction of **Old Products** gets higher, the importance of R&D gets higher and it gets larger values to add more products to the product portfolio. The more **R&D-Quality Budget** we get, the bigger the **R&D-Quality Budget Ratio** gets which is basically the division of the current budget with the previous quarter's budget. This adds to the **Quality Inflow** more.

As **Quality Inflow** increases, **Quality** increases and there occurs more **Quality Outflow**. Then this affects **Quality** and lowers its value (Loop B4). The more **Quality Outflow**, the more **Estimated Quality Outflow** takes place and the more **Quality Inflow** gets (Loop R3). This mechanism works like an adjustment for the quality. The managers are replacing the depreciated quality with the new amount.

There is also the concept of the **Desired Quality** which the company seeks to attain. The difference between this value and the actual one gives us the **Quality Discrepancy** which would add more to the **Quality Inflow** to adjust the quality toward the desired value (Loop B5).

After calculating the **Quality Balance,** which is the ratio of current **Quality** with the **Desired Quality**, this affects the **R&D-Quality Budget**. **R&D Budget** is a general budget, which is composed of two subbudgets, namely **R&D-Quality Budget** and **R&D-New Product Budget**. If we get closer to the **Desired Quality,** then there would be less budget allocated next time around. Finally, we come up with the new budget for **R&D-Quality** where we have started (Loop R4 and B6) (Fig. 4).

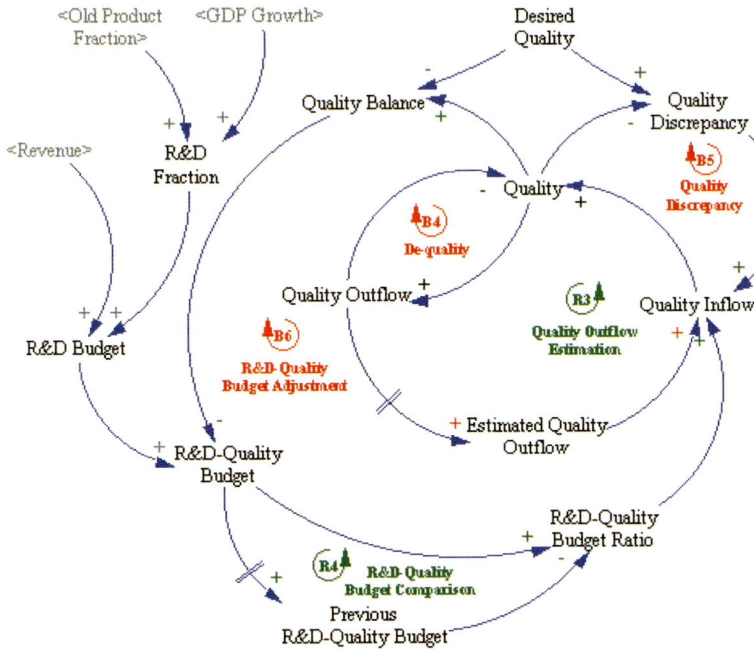

Figure 4: Causal loop diagram of the R&D-quality sector.

R&D-New product sector

One would remember that the **R&D-New Product Budget** is dependent on the **R&D-Quality Budget**. Quality budget is subtracted from the total R&D budget to get the product development budget. By dividing this budget with the value or the cost of a new product, we get the **Launching Rate** of new products. The more **New Products** we get, the more obsolete they become.

In this sector, one can see that there is a stream of products from **New Products** to **Mature Products** and then to **Old Products**. It aims to reflect the "Product Life Cycle" concept in the model. As the **New Products** are increased there would be more transition to the **Mature Products** and then to **Old**

Products. Each phase has its own obsolescence time (Loops B7, B8, and B9). This approach includes imitation by competitors implicitly.

Total Products is the sum of all three stages described above, so it is positively affected by these products. We compare this number with a **Reference Number of Products** available in the market. The target here is to see if the company is lagging the market in terms of product portfolio. The ratio is called **Product Balance Ratio** which is used in the next sector. In addition to this availability issue, the products' being too old also is a threat for the company in the competitive arena. If the fraction of **Old Products** is increased, the company should allocate more **R&D Budget** and if this ratio is high, **Customers** would begin attrition from the company.

The fractions of the products are also calculated so that they can be used for the Customer Sector. They are used for **Price** calculation.

Customer sector

This sector may said to be the main engine of the model due to its importance for the Customer base. As Warren (2002) also states **Customers** are the main and crucial stock of the company.

As it was stated earlier, there is a **Customer** base in the market active in transactions with the company and there is also another pool of customers who are called **Potential Customers**. As the **Potential Customers** are increased, there would be more **Adoption Rate** to the actual **Customers.**

The actual **Customers** may influence the **Potential Customers** to adopt the company's products. This happens in the following way: As the number of **Customers** increases, there would be a certain number of contacts that they could do to affect by taking the **Contact Rate** into account. The more **Customer Contacts**, the more **Possible Contacts between Potentials and Customers**. However, not all of these contacts are persuasive for the Potentials. In other words, there is a **WoM Probability** that this contact would turn into a successful one for the company. The bigger this probability gets, the bigger adoptions would occur. This process is called **Word-of-Mouth** effect (Loops R6 and R7).

On the other side, the company also has many other adoptions from different features of the product or the company. There are also **Adoptions from R&D-Quality** (Loop B14), **Adoptions from R&D-New Products** (Loop B13), and **Adoptions from Marketing** (Loop B15). They affect the **Adoption Rate** in the same direction.

A customer may give up the product permanently or temporarily. He may be added to the Potentials pool or may be drawn out of the system (Loops B11, B10, and R5).

Potential Customers are increased by **Potential Growth Rate** which is one of the inflows of the stock. **GDP Growth** also affects the **Price**. **GDP Growth** and **Indicated Price** are multiplied basically with an assumption. Then each **Customer** is believed to bring some revenue to the company. The more **Customers** and **Price**, the more **Revenue** there occurs.

There are also some major loops that are important. As the causal loop diagram is composed of four sectors, it may be difficult to visualize that **Revenue** is allocated to Marketing and R&D, roughly speaking. These two departments get **Customers** to the company which would bring more **Revenue** next time around. This structure may be easily visualized by the help of Fig. 5 which is the Sector relation figure (Fig. 6).

Stock-Flow Diagram

In the previous section, the causalities between the variables of the model were explained briefly and qualitatively. Readers interested in a more detailed analysis of the model in terms of understanding the underlying structure may contact the authors to request the list of the model equations as well as the Stock and Flow figures.

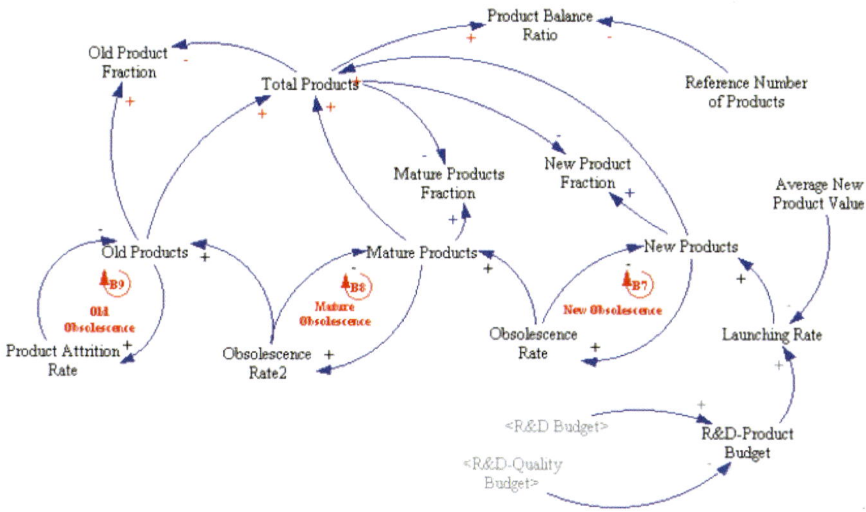

Figure 5: Causal loop diagram of the R&D-new product sector.

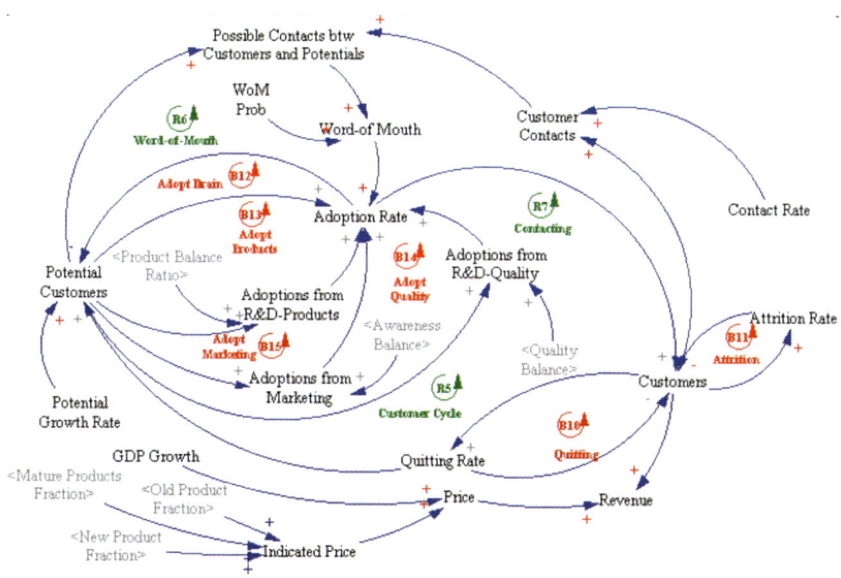

Figure 6: Causal loop diagram of the customer sector.

Selection of stocks

As Sterman (2000) suggests, the snapshot test is used for the selection of the stocks. The main stocks after this analysis are as follows:

- Customers
- Potential Customers
- Product Quality
- Customer Awareness
- New Products
- Mature Products
- Old Products
- Electro Market

These stocks are mentioned in the following sections. The time unit of the model is month and the explanations below are done based on this statement. They are also the resources of the company from the point of view of Warren (2002).

These structures were more or less explained in the Causal Loop Diagrams before, but for space saving purposes it was decided not to illustrate all equations. Therefore, for the complete stock-flow equations and for input functions, interested readers may contact the authors. The equations described in the text belong to Base Run.

Testing and Validation

In testing a model to see if it is an adequate representation of the reality, Barlas (1996) proposed a sequence of validation tests for formal validation. In this section, this sequence will be followed with the chosen and relevant tests for the model. As the model does not take raw and actual data into account, there is no attempt to replicate the history. All tests are concerned on the structure and sensitivity.

Direct Structure Test

These tests assess the validity of the model structure by direct comparison about real system structure. One will not be able to see a simulation, but rather structure and parameter confirmation with dimensional consistency.

In this part, each equation is considered alone and evaluated with the current knowledge in the literature. This was in fact done with the stock and flow model building at the same time. Each equation was checked and evaluated if it reflects real life. Zero fudge-factoring is aimed at and the units of each side of the equations are checked. For instance while a "stock" has the unit of *item,* a "rate" has the unit of *item/time* since they reflect flows in the system.

Structure-Oriented Behavior Tests

These tests are rather concentrate on testing the behavior of the model with the simulations carried out. One compares the expected mode of behavior with that of the model generated. Then one concludes that the model is either good enough or needs further work and reverse engineering.

Extreme conditions test

This test includes simulation through time by giving extreme values to the selected variable and comparing the model generated behavior to the anticipated behavior of the real system under the same extreme condition (Barlas, 1996).

Price

Price is set at 0 by making

$$\text{Price}_{\text{Mature}} = \text{Price}_{\text{New}} = \text{Price}_{\text{Old}} = 0 \qquad (1)$$

Price is chosen to be very close to zero which is 0.000000001 in order to eliminate the divisibility by zero error. In addition due to this effect, simulation time has been shifted back to-48 because even this budget affects **Quality** and **Awareness**, so it is filtered for the reader for better understanding the extreme condition (see Fig. 7 for the results and Run2).

Expectedly, when **Price** goes to zero, **Revenue** goes to zero and as a result there occurs no budget for the departments of the firm for both marketing and R&D. Then the related measures such as **Product Quality** (bottom left), **Customer Awareness** (bottom left), and Product variety (bottom right) get closer to zero. They are not zero yet at the end of the simulation for 120 months, but it is assured that they go to zero for long-period simulations. After these effects Customers (top left) begin to quit the company due to lack of prementioned measures and so the **Market Share** converges to zero. In addition the rates related with the **Customers** and **Potential Customers** go to zero (bottom left) which is the equilibrium for the system. The company collapses.

The behavior explanation of the model would be done in the next part, so the reason behind the peaks for the **Product Quality** and **Customer Awareness** would be investigated later.

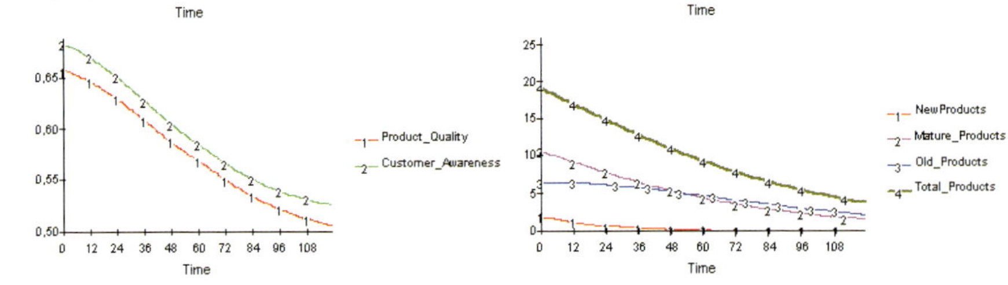

Figure 7: Main stocks of the model after price = 0.

- Price is set at Maximum by making

$$\text{Price}_{\text{Mature}} = \text{Price}_{\text{New}} = \text{Price}_{\text{Old}} = 100 \tag{2}$$

which is a relatively high value (see Fig. 8 for the results and Run3).

When **Price** goes to 100 which is a relatively high value, **Revenue** goes to a high value in the beginning due to high **Customer** level and as a result there occurs high budget for the departments of the firm for both marketing and R&D. Then the related measures such as **Product Quality** (top right) and **Customer Awareness** (top right) experience goal seeking growth. **Product variety** (bottom right) goes to higher values also. As it is a third order structure, after the first stock, this response is reflected to the others. As the **Market Share** is the ratio of **Customers** to **Total Customers**, until time 60, they gain half of the market which is evident from the figure on top row of Fig. 8.

However, **Potential Growth Rate** of **Potentials** increases continuously and the net increase in the **Customers** is less than that of the **Potentials**. This may really be the case in reality

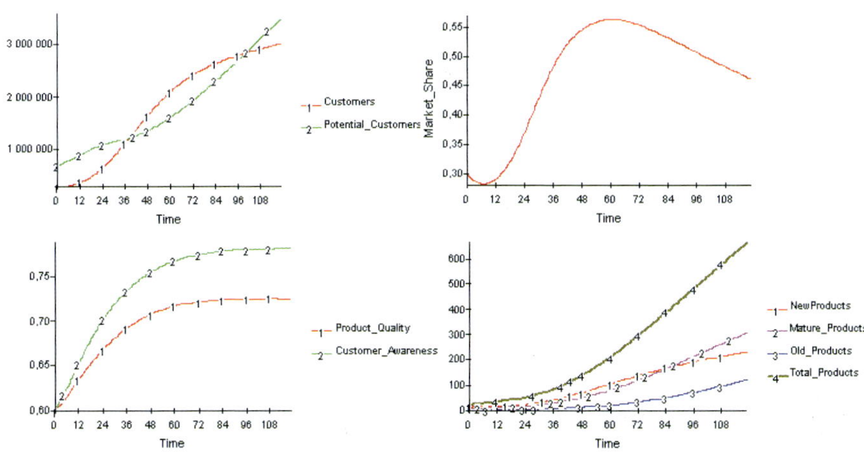

Figure 8: Main stocks of the model after price = 100.

because after a time when a company gets a large customer base, it shows basically that it is a profitable area. This is an indicator for most of the entrants to get into business there. Therefore, the company has to share this customer base with the rivals. In addition, as the price is on the high extreme, the **Combined Effect** — that affects the **Potential Growth Rate** — really takes high values so the **Potential Customers** reach higher values earlier than it would have got otherwise. This may be the interpretation of the loss of **Market Share**.

Average new product value (ANPV)

- **ANPV is set at 4545 which is a relatively low value (see** Fig. 9 **and Run4)**

When ANPV goes to 4545 which is a relatively low value, there can be made more new products next time around and the transition rates from the New Products to the other aging products, go more each time. They grow continuously (bottom right). This brings an "early" success in the Product Balance which would affect both Product Quality and Customer Awareness in the same way (bottom left). The net increase in

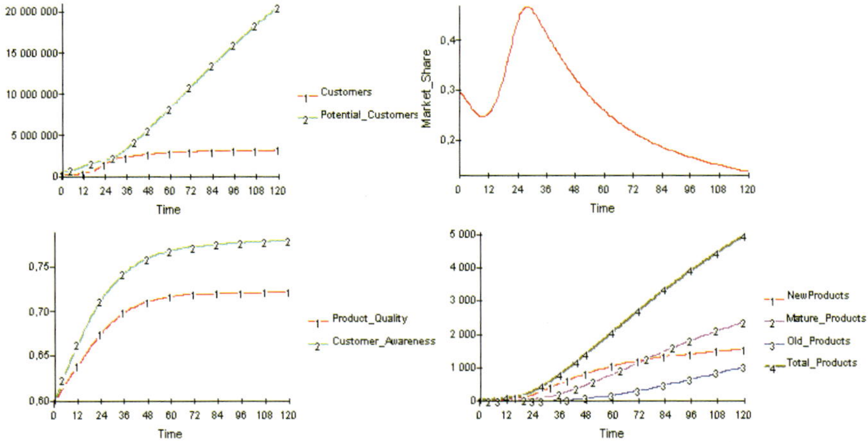

Figure 9: Main stocks of the model after ANPV = 4545.

the Potential Customers is again bigger than that of Customers (top left) and the Market Share drops drastically after time 30, because the number of Customers stays very low according to the Total Customers (top right).

- **ANPV is set at 1000,000 which is a relatively high value (cf.** Fig. 14 **and Run5)**

When ANPV goes to 1000,000 which is a relatively high value, there can be made less new products next time around and the transition rates from the New Products to the other aging products, go less each time except for Mature Products until time 24. They go down to zero (bottom right) due to high product development costs. This brings an unsuccessfulness in the Product Balance which would affect both Product Quality and Customer Awareness in the same way after a time (bottom left). This does not affect them immediately because they get their budgets from the company, but after a time new product unsuccessfulness brings a disease to whole company. As a result of this, Customers quit the company (top left) and the Market Share goes really high until time 50, but drops down later on (cf. Fig. 10).

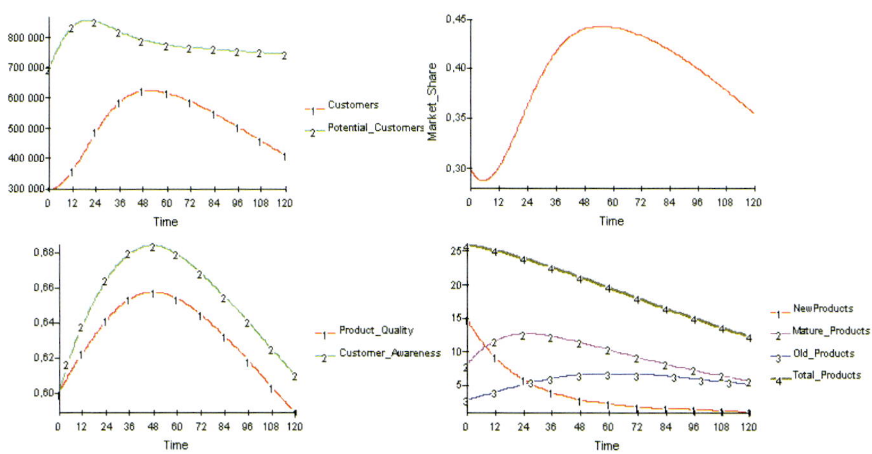

Figure 10: Main stocks of the model after ANPV = 100,000.

Reference R&D fraction (RRDF) and reference MB fraction (RMBF)

- **They are both set to 0.001 which is a relatively low value.** It is not 0 in order to overcome the problem of divisibility by zero in the model. Simulation is started from 48 to 120 in order to eliminate this divisibility by zero effect (see Fig. 11 and Run6).

When **Budget Fractions** go to 0.001 which is a relatively low value, there can be made less (almost zero) and less budget next time around. As the budgets go lower, less **New Products** are generated and the transition rates get low (bottom right). The budgets for the quality and awareness get decreasingly increasing values and this reflects to the **Awareness** and **Quality** (bottom left). As a result of this, **Customers** quit the company (top left) and the **Market Share** goes to zero (top right).

- They are both set to

$$\text{RRDF} = 0.5, \text{RMBF} = 0.5 \tag{3}$$

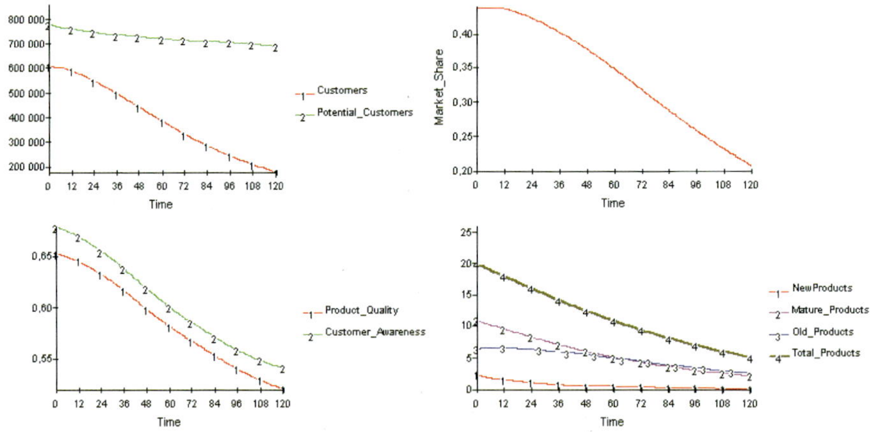

Figure 11: Main stocks of the model after Reference Fractions = RRDF = RMBF = 0.001.

The reason why they were given 0.5 is that it was presumed that there were just two departments in the company for the extreme case (see Fig. 12 and Run7).

When Budget Fractions go to 0.5 which is the maximum meaningful value, there can be more budget next time around. As the budgets go higher, more **New Products** are generated and the other product type stocks go higher (bottom right). The budgets for the quality and awareness get decreasingly increasing values and this reflects to the **Awareness** and **Quality** (bottom left). As a result of this, more **Customers** adopt the products of the company decreasingly (top left). This can be seen in the **Market Share** figure. The net increase in the **Customers** cannot keep pace with the **Potential Customers** (top right). As a result **Marketing Share** starts dropping at the point where the net increase in **Customers** is equal to the net increase in **Potential Customers**.

Sensitivity test

This test consists of determining those parameters to which the model is highly sensitive, and asking if the real system would

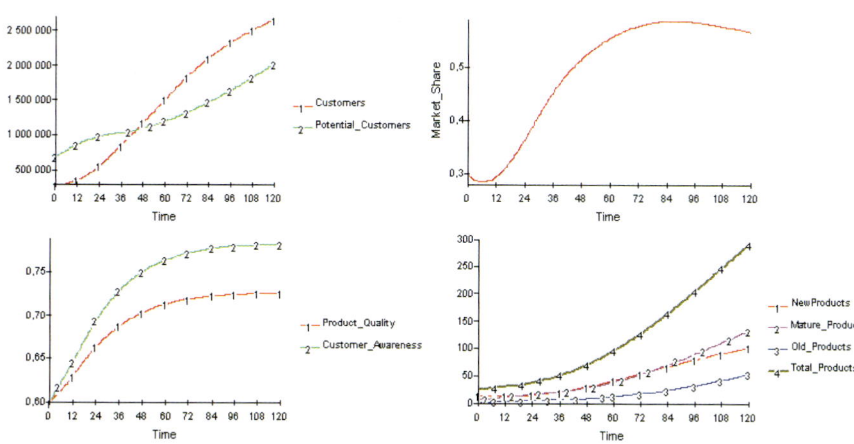

Figure 12: Main stocks of the model after Reference Fractions RMBF = RRDF = 0.5.

exhibit similar high sensitivity to the corresponding parameters (Barlas, 1996).

Contact rate

It is common sense that no company can estimate **Contact Rate** exactly because we are human beings and we have desires and changing goals each time. This makes us different from animals and plants. We behave in a random way most of the time. While wandering through the city we coincide with friends or there are lots of phone calls between each side. It is really impossible to say that I have made five Contacts this month, because it is even arguable what a contact is and what its effect is on the phone or face to face. There is a huge uncertainty in this rate, so one must pay special attention to these kinds of variables in a model which bring uncertainty with itself. This was the departure point for selecting this variable as a sensitivity variable. One can see the result of the analysis in Fig. 13 and Base Run and Run 9, 10, and 11.

As the **Contact Rate** is increased, one can realize that in the initial 6 months or so, the behavior in the **Customer** stock shows a different trend than the base case where the rate was two person/month/person.

It is obvious that if the **Contact Rate** is increased, there would be more space for potential growth in the **Customers** via WoM (Word-of-Mouth) the next time around. However, the trend that **Marketing Share** and **Customer** follow is different each time.

The company should assume high **Contact Rates** if they had done advertising or distributed free products previously in the market so that there occurs more **Customers** interested in making WoM effects to others. The reverse case is that if the company is a start up without previous efforts, then a low **Contact Rate** should be chosen. The middle values can be chosen as an insight for the model for the other types of mature companies.

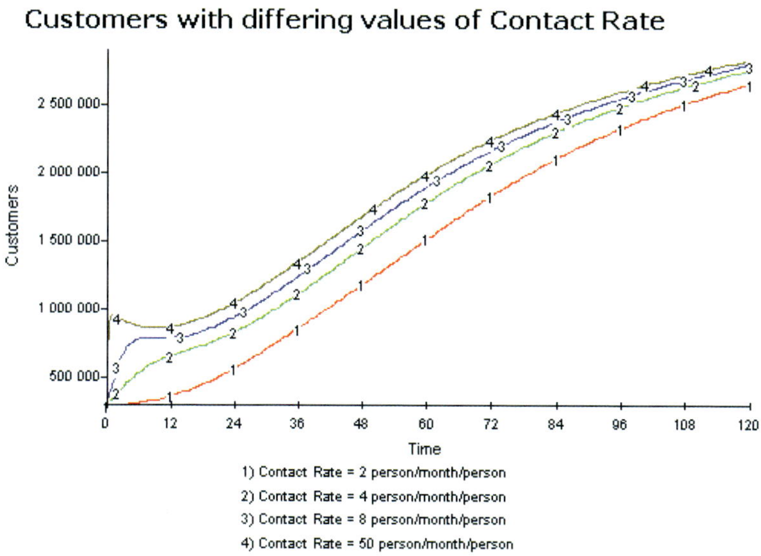

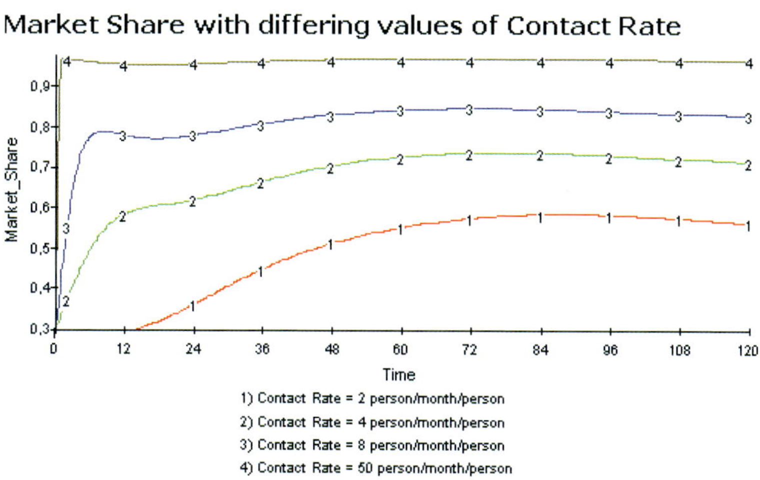

Figure 13: Market share and customers with differing values of contact rate.

Simulations and Behavior Discussions

In this section, it is aimed to express why the model generates what it generates. The discussion would be made with the Base

Run. The stocks would be explained with the underlying **Net Flow** to the stock. Net Flows are calculated instantly such as:

$$\text{Net Flow}_S = (\sum \text{Inflow}_S) - (\sum \text{Outflow}_S) \quad \text{where S is the index of the stock} \tag{4}$$

Base Run

In this run, the GDP Growth and WoM Probability are designed in the following way:

$$\text{WoM\_Prob} = \text{NORMAL } (0.1, 0.005, 2) \times 0 + 0.1/12 \times 1 \tag{5}$$

$$\text{GDP\_GROWTH} = (\text{SINWAVE } [\text{RANDOM } \{0.01, 0.05, 6\}, 48] \times 100/12) \times 0 + (0.1 \times 1/12) \times 1 \tag{6}$$

Potential customers

One can see the Net Flow formula for the Potential Customers below:

$$\text{Net Flow}_{\text{Potential Customers}} = (\text{Potential Growth Rate} + \text{Quitting Rate}) - (\text{Adoption Rate}) \tag{7}$$

As one can see in the base run, Potential Customers increase throughout the simulation. Around time 33, the growth is the minimum which can easily be seen from the Net Flow graph. Until that time Net Flow decreases to a certain amount around 3000, but this does not mean that stock would do the same because a stock is the result of the integration of its Net Flow with time plus whatever it had previously. In other words, the area under the Netflow graph plotted against time gives us the additional stock value. During this period Potential Customers follow a goal seeking behavior with growth. After time 33, Net Flow starts increasing like a goal seeking behavior, but not a pure one. The stock, however, follows a transition in the growth trend from linear to some form of growth and then gets closer to linear growth again (Fig. 14).

Customers

One can see the Net Flow formula for the **Customer**s below:

$$\text{Net Flow}_{\text{Customers}} = (\text{Adoption Rate}) - (\text{Quitting Rate} + \text{Attrition Rate}) \quad (8)$$

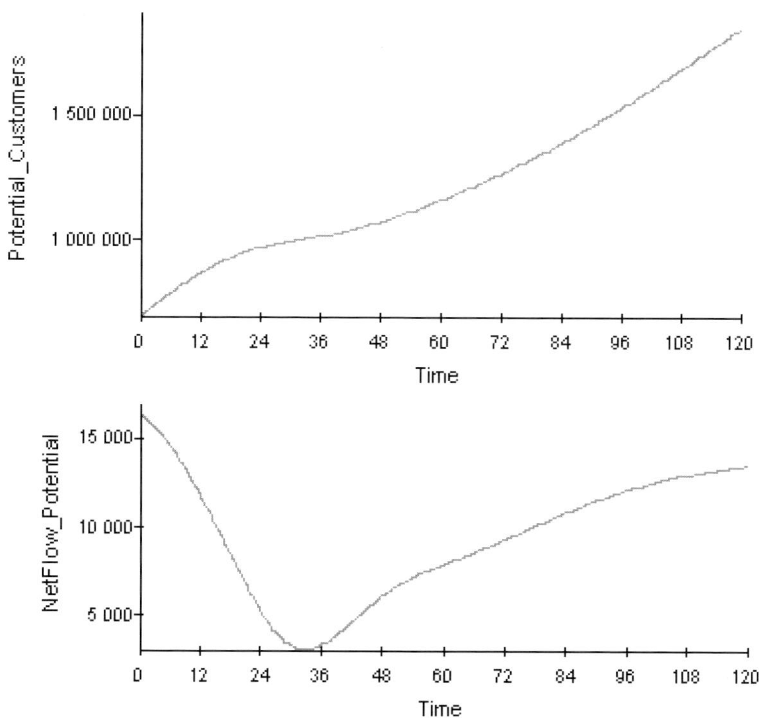

Figure 14: Potential customers and its net flow.

Customers show an increasing trend with different characters throughout the simulation. It first increases like an exponential growth due to its net flow's almost linear increase till time 24. Then the Net Flow decelerates and comes to a smooth valley during which **Customer**s increase very close to a linear way. After time 66, Net Flow starts to drop down which would bring a decreasing increase to the **Customers** (Fig. 15).

Product quality

Product Quality shows rather a distinct type of behavior. Until time 12, there is a growth close to exponential growth in the **Netflow** and this reflects to the stock behavior certainly. Stock almost behaves in the same manner, but suddenly at time 12,

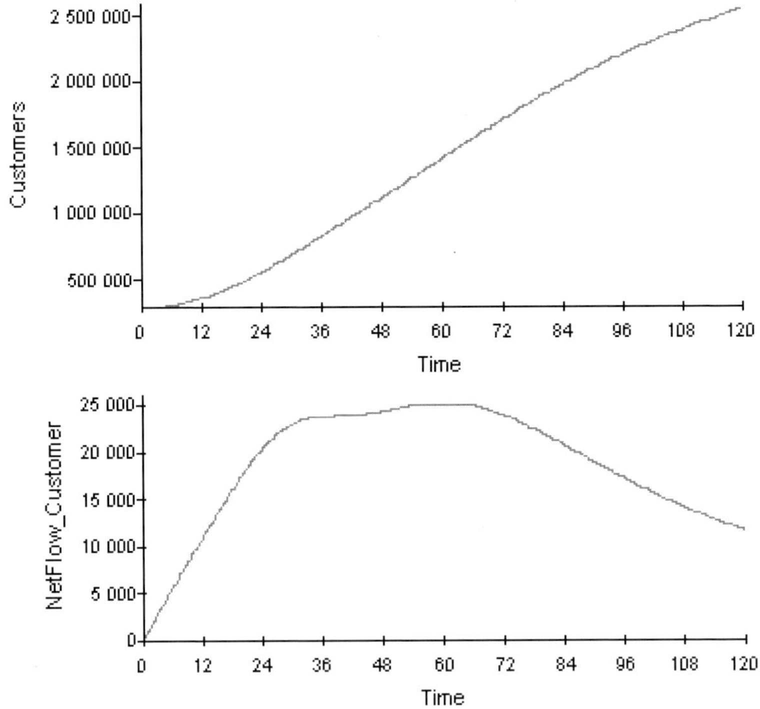

Figure 15: Customer and its net flow.

Net Flow starts decreasing till around 18 and after that there is a drastic decrease. Between time 12 and 18 we cannot see a specific changing behavior in the stock due to the impact's effect is valid for a short time and net flow's being small. The area under that time range is really small to change the stock so that we can see a differing path. (We could have seen the behavior if we just make the graph bigger in scale which is not feasible here.) After time 18, Product Quality follows a path to reach its goal, but stabilizes below that value in the long term (Fig. 16).

One can see the 3D phase plots of the Product Quality (*z* axis) and its two flows, namely Quality Inflow (*x* axis), and

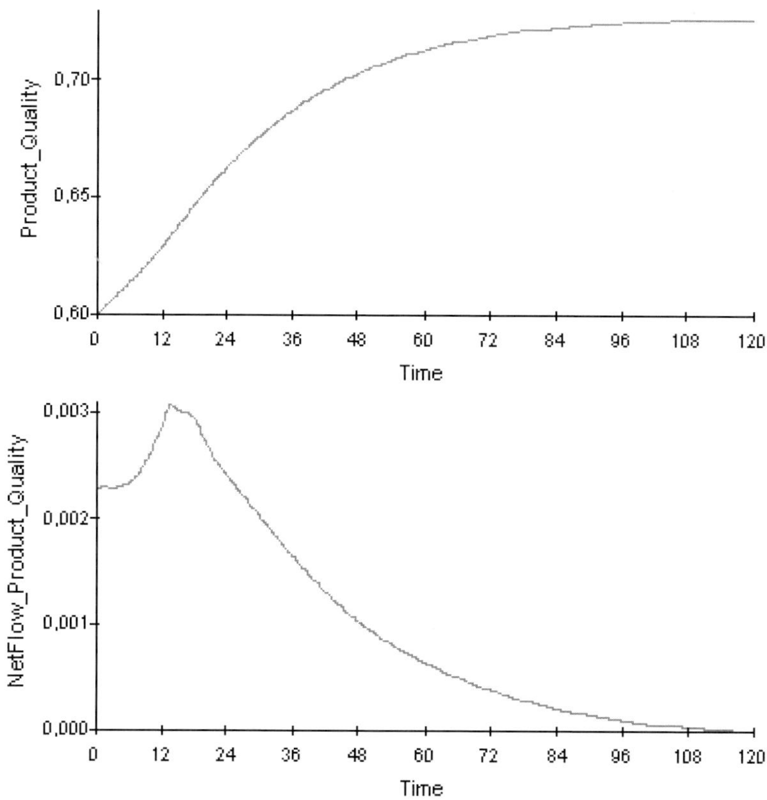

Figure 16: Product quality and its net flow.

Quality Outflow (*y* axis). Hot colors such as red and its neighboring colors reflect higher values (Fig. 17).

Customer awareness

Customer Awareness also shows a similar behavior with the **Product Quality** stock. However, the behavior of the **Net Flow** is not the same with the **Product Quality**. Until time 7, **Net Flow** decreases decreasingly down to a level and gets a local minimum there. During this time **Customer Awareness** increases but never gets a flat behavior because the **Net Flow** is still a positive value, not zero. Then from time 7 to 12, **Net Flow** starts increasing which affects the stock the same way. The exact behavior related to those **Net Flow** changes cannot be seen due to the already mentioned reasons of scaling. Then from time 12–18, the rate of increase in the **Net Flow** just slows down and expectedly stock has still its growth over time. After time 18, **Net Flow** starts going down which would affect the stock with the goal seeking type of behavior due to the **Net Flow**'s decreasing to zero which means there is no growth at all (Fig. 18).

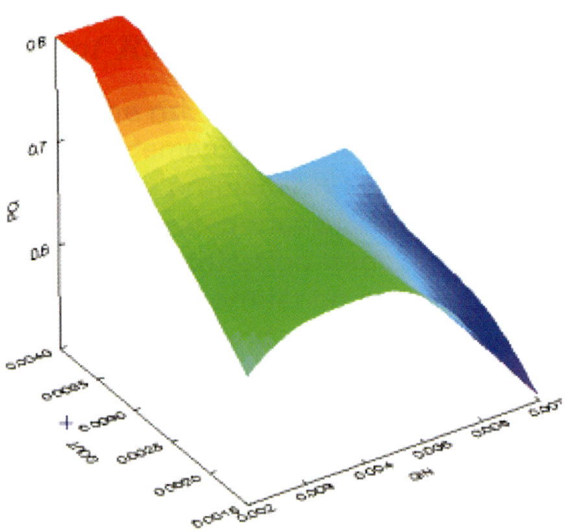

Figure 17: Product quality and its flows.

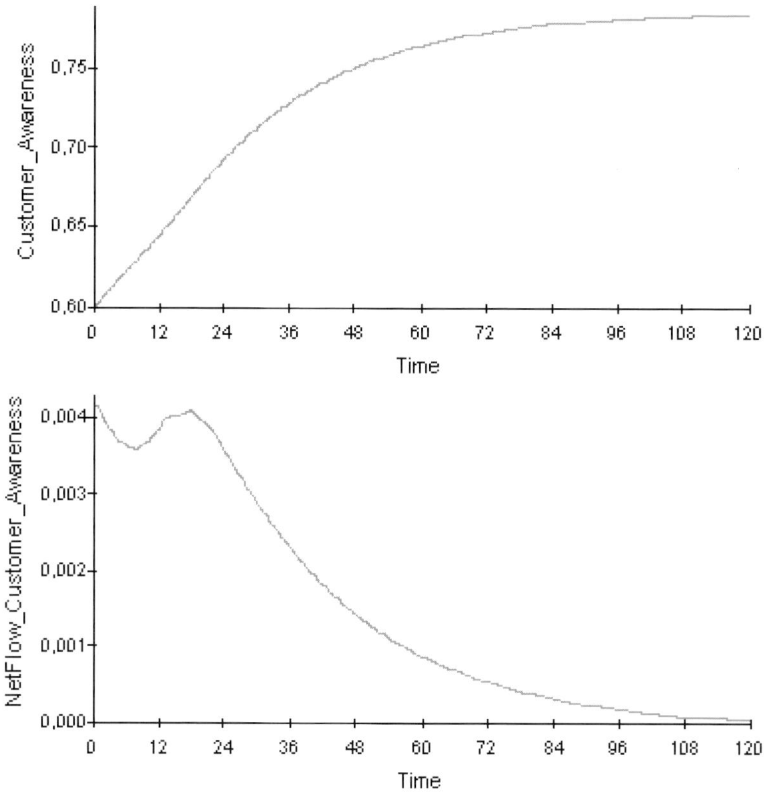

Figure 18: Customer awareness and its net flow.

One can see the 3D phase plots of the **Customer Awareness** (*z* axis) and its two flows, namely **Making Aware Rate** (*x* axis), and **Forgetting Rate** (*y* axis). Hot colors such as red and its neighboring colors reflect higher values (Fig. 19).

New products

This stock is the first in a series of aging products. Although Net Flow increases until time 24 like an exponential function, one should pay special attention to the values it takes, because until time 12 it takes negative values which increase (get closer to zero) through the mentioned time range. **New**

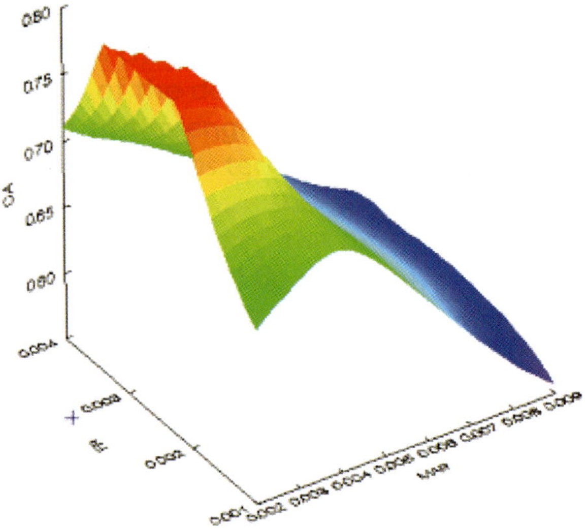

Figure 19: Customer awareness and its flows.

Products initially decrease very little due to this reason. In the time range between 12 and 24 the increase in the Net Flow still goes on and the behavior in the stock is exponential growth. However, there is an inflection point for the Net Flow which is very close to time 24. Net flows start to increase decreasingly which would affect the **New Products** get closer to the linear growth from exponential growth. After time 78, Net Flows start to go down which would affect the stock get closer to a goal value (Fig. 20).

Mature products

Net Flow for the **Mature Products** follows a negative goal seeking behavior until time 18, but still has a positive value. This would affect the stock, **Mature Products**, gets very close to a flat (but not) region around the same time 18. After time 18, Net flow starts to increase and experiences an inflection in the growth at about 55. During this time period (18–55) stock follows an exponential growth. After that time the stock behavior

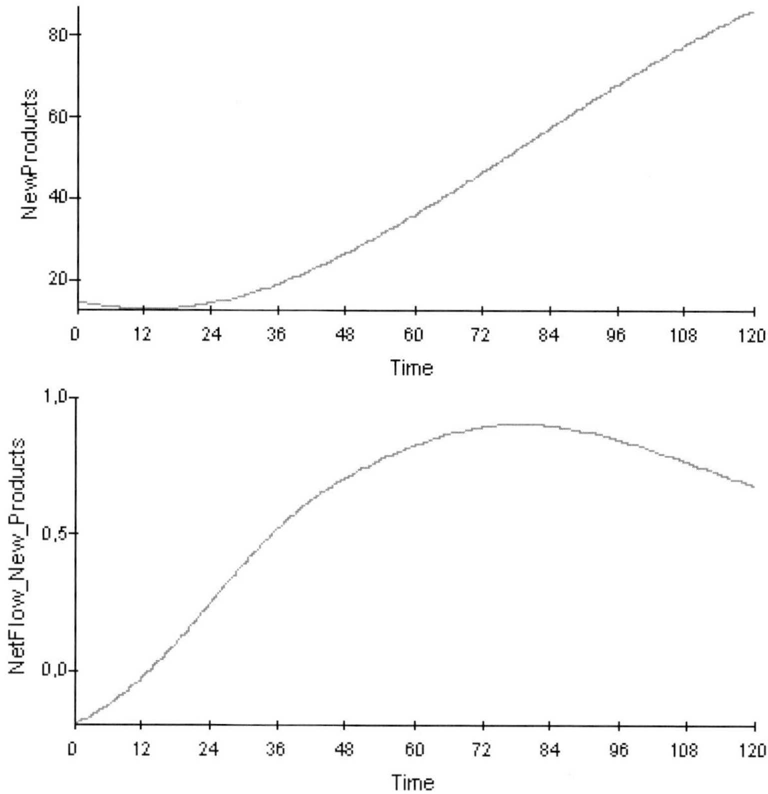

Figure 20: New products and its net flow.

gets closer to a linear growth due the fact that Net Flow starts to stabilize at around a value of 1.5 (Fig. 21).

Old products

This is the third stock in the series of product stream (in the New Product Sector). The Net Flow of the **Old Products** is a similar version to that of the **Mature Products** except there is decrease in the Net Flow of the **Mature Products** initially. This is because of the fact that, in the higher orders of the aging chain, the last stocks just follow the previous ones and the previous ones follow the first stock. Taking all time constants

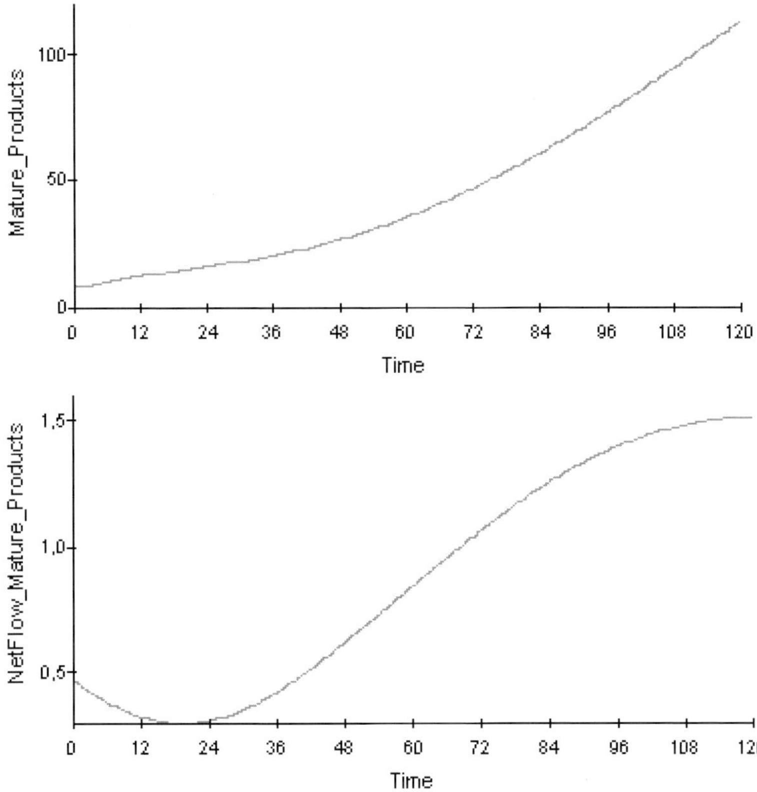

Figure 21: Mature products and its net flow.

and delays into account, the last stock just does not respond the same way to the system as the first stock does. We can see this in the Net Flow behavior of the **Old Products** and **Mature Products**. What matters are the ratios of the adjustment times in this aging chain. This defines the behavior of the product chain members. As the Net flow, here, increases continuously which would affect the stock the same way. Net Flow gets a stable value much after the simulation, then the stock would increase linearly or exponentially depending on the options for the Net Flow (Fig. 22).

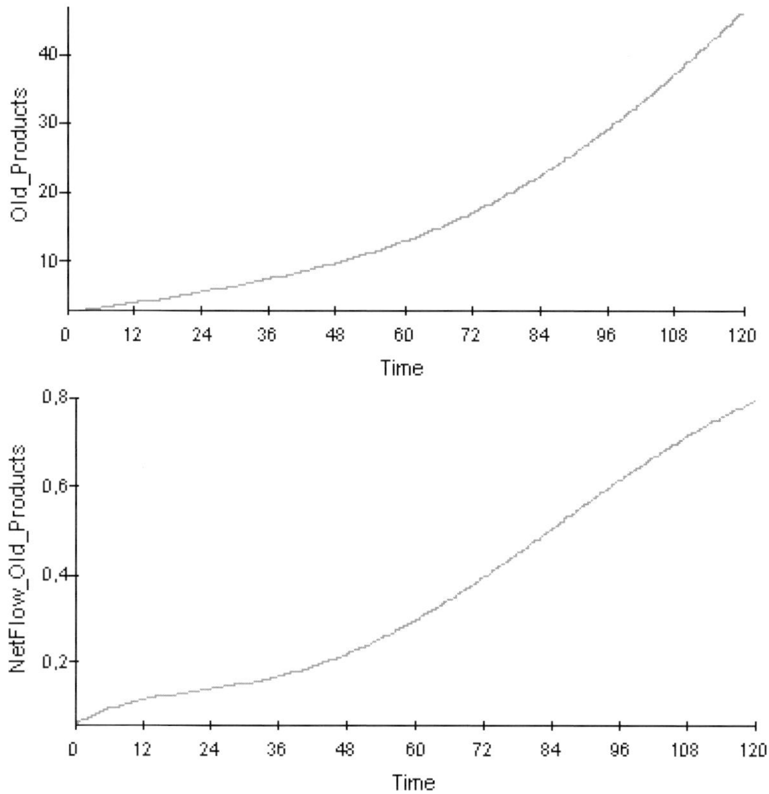

Figure 22: Old products and its net flow.

Alternative Run

In this run, just the GDP and WoM Probability are changed in the following way:

$$\text{GDP\_GROWTH} = (\text{SINWAVE}\,[\text{RANDOM}\,\{0.01, 0.05, 6\}), 60]/12) \quad (9)$$

GDP is given with a sinus wave whose amplitude is a random function between 0.01 and 0.05 with period 60 months

(5 years). It is assumed that GDP of the country oscillates (due to historical data) between certain values. These values are divided by 12 just for the input function and monthly adjustment.

$$\text{WoM Probability} = \text{NORMAL}\ (0.1{,}0.005{,}2) \qquad (10)$$

A distribution is given to the probability for the Word-of-Mouth effect. The probability is reduced by 100 assuming that it is not so probable for a contact to end up with an actual Customer.

One can see the summarized results of the alternative simulation below: (Alternative Run) (Fig. 23).

The oscillating level of GDP Growth (top right) is averaged for R&D and Marketing (middle right). In other words "*White Noise*" has been utilized for random input to GDP growth. However, as in reality, these ups and downs are in a way

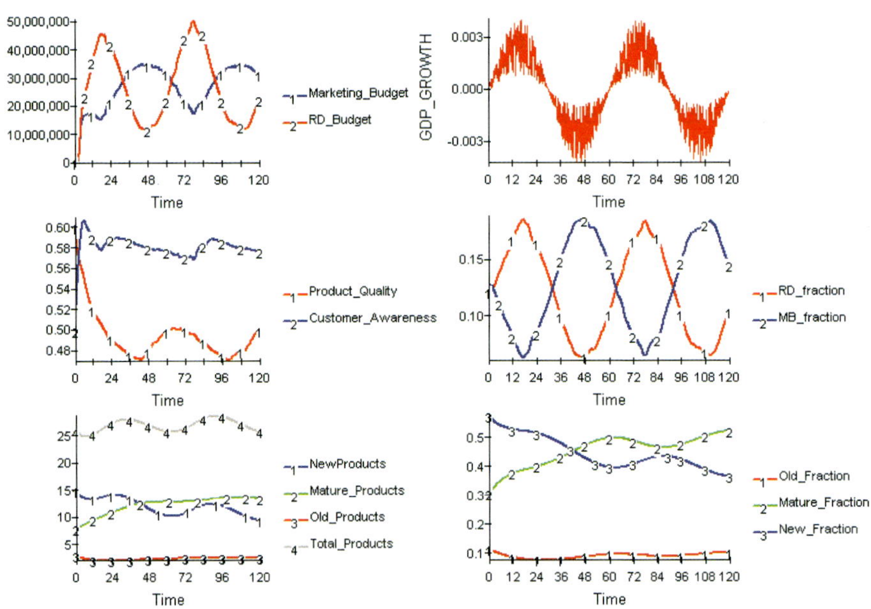

Figure 23: Some variable results for alternative run.

smoothed via "*Pink Noise*" which addresses the fact that our perception of a real issue comes with a smoothing effect. We have utilized the heuristic called "anchor-and-adjustment". We anchor to an initial value and adjust it through a perception process.

GDP Growth in turn affects the budget amounts much (top left). **Marketing Budget** and **R&D Budget** are oscillating with an increasing trend. Both **Product Quality** and **Customer Awareness** get oscillating values (middle left). In addition **Mature Products** also get higher values throughout simulation (bottom left). The fractions according to these trends in the products are seen on the bottom right. **New Fraction** and **Mature Fraction** dominate the spectrum and there is an oscillating trend in the last 80 months and **Old Fraction** is always around 0.1 level (Fig. 24).

Market Share skyrockets drastically (top right) in the beginning and after a time it stabilizes with very minor oscillations around a certain value. This can be interpreted as the market is cannibalized as a consequence of the allocation strategy. This is due to the timing of R&D and Marketing activities. On the other hand, being a crucial component of the **Revenue**, **Price** is

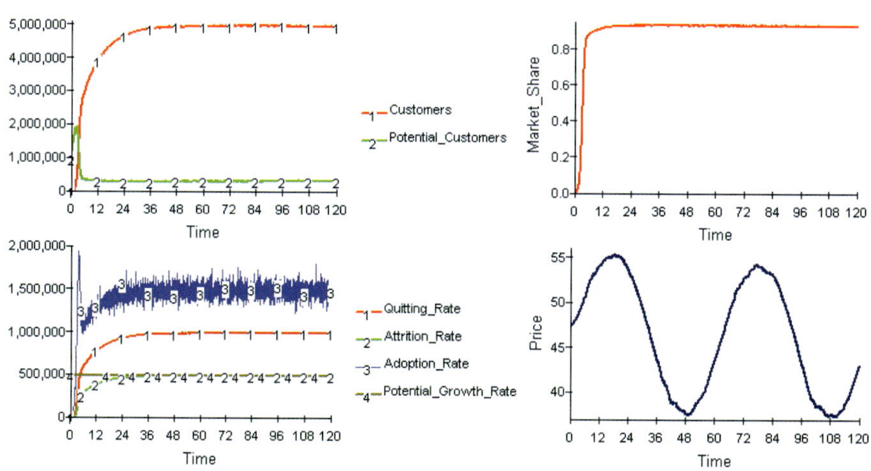

Figure 24: Customer and market share figures.

kept low during the upturns and high during the downturns. It is oscillating (bottom right) mostly due to the **GDP effect on Price**. The economic developments and price trends might seem contradictory during the downturns especially, but this effect is also compensated during the upturns when the incomes of population are relatively higher. Beyond compensating, the benefits during the upturns are much bigger than that of the downturns, because the market share is kept at a high level. One other factor is obtaining the market share which is an indicator also for future profits in a way.

The objective of this alternative run was to show the effect of budget allocation between marketing and R&D activities by taking into consideration the GDP growth of the economic environment. The main hypothesis beyond this run is that companies should invest in R&D activities when the economic trend is upturn, since there is enough source to fund them. However, when the economic situation is downturn, companies should try to focus on their marketing activities to boost their sales capacity.

Policy Design and Evaluation

In this section, we aim to design some policy tests for the Base Run which has the simple assumptions. The departure point for the design of the policies was the decision for the Marketing Spending and R&D Spending. It was assumed that the share to the total of R&D and Marketing Departments is a fixed percentage of the revenue which is 25%.

Before going on with the policies, it would be beneficial to discuss survey results for allocation of the budgets between marketing and R&D for high, medium, and low tech companies. The survey was carried out online on a Yahoo Groups poll site which is devoted to strategy development group in Turkey. The results seem to confirm Trott (2002)'s classification.

High-Tech Case Results: 68% of the participants gave more weight to R&D than marketing. This shows a tendency in this sector to promote innovations more than marketing activities. However, the participants were indifferent between 10% and 15% (M-R&D) allocation and 5%–20% allocation. This gives flexibility to the policy design for these types of companies.

Medium-Tech Case Results: 80% of the participants gave more weight to marketing than R&D. This shows a tendency in this sector to promote marketing activities more than innovations. The participants were clear about the allocation since more than half of them decided on 15%–10% allocation. Extreme R&D allocation is very low.

Low-Tech Case Results: 83% of the participants gave more weight to marketing than R&D. The participants were clear about the allocation since half of them decided on 20%–5% allocation. There are no extreme R&D allocations (i.e., 20% R&D, or 24% R&D).

Policies

Policy 1 — base policy

Policy 1 is thought to be the base case for an average electronics company who invest both in R&D and Marketing. The percentages are as follows: Please see Run P1.

$$\text{Marketing Percentage} = 20\% \tag{11}$$

$$\text{R\%D Percentage} = 5\% \tag{12}$$

Policy 2: marketing emphasis

Policy 2 is thought to be the base case for an average electronics company who invest less in R&D and more on Marketing. The percentages are as follows: Please see Run P2.

$$\text{Marketing Percentage} = 24\% \tag{13}$$

$$\text{R\%D Percentage} = 1\% \tag{14}$$

Policy 3: R&D emphasis

Policy 3 is thought to be the base case for an average electronics company who invest more in R&D and less on Marketing. The percentages are as follows: Please see Run P3.

$$\text{Marketing Percentage} = 5\% \tag{15}$$

$$\text{R\%D Percentage} = 20\% \tag{16}$$

Policy Evaluation

Policies designed above are evaluated in this section. First of all, a manager must face the reality that he has to choose among some policies to implement and by doing so he has to do some trade-offs. They are clearly and intuitively realized in the results of the runs for the policy parameters below. The representations of the policies have the same number indices in the graphs: Curve 1 & Policy 1, Curve 2 & Policy 2, and Curve 3 & Policy 3 (Fig. 25).

In terms of **Customers,** Policy 3 seems to be the best policy, because heavy investment of R&D in Quality and New Products has a dominating effect on all of the departments of the company. Policy 2 is the worst which can give an insight to the manager that even though the **Marketing Spending** is very high, the return is not so as it was expected to behave accordingly. Low R&D investments can affect the marketing in a decreasing way. Policy 1 is the mediocre among these policies.

In terms of **Quality** and **Awareness**, Policy 3 seems to work best among the other policies, because the company can reach a higher value earlier than the other policies. Policy 2 is again the worst policy to adopt under these circumstances. The achievement of the same level according to the other policies is much later. Policy 1 is still the second choice.

In terms of **Price**, Policy 3 is still the best. It is much over the other two policies till time around 90. This is good because the revenue is made up of **Customers** and **Price** mainly. During this

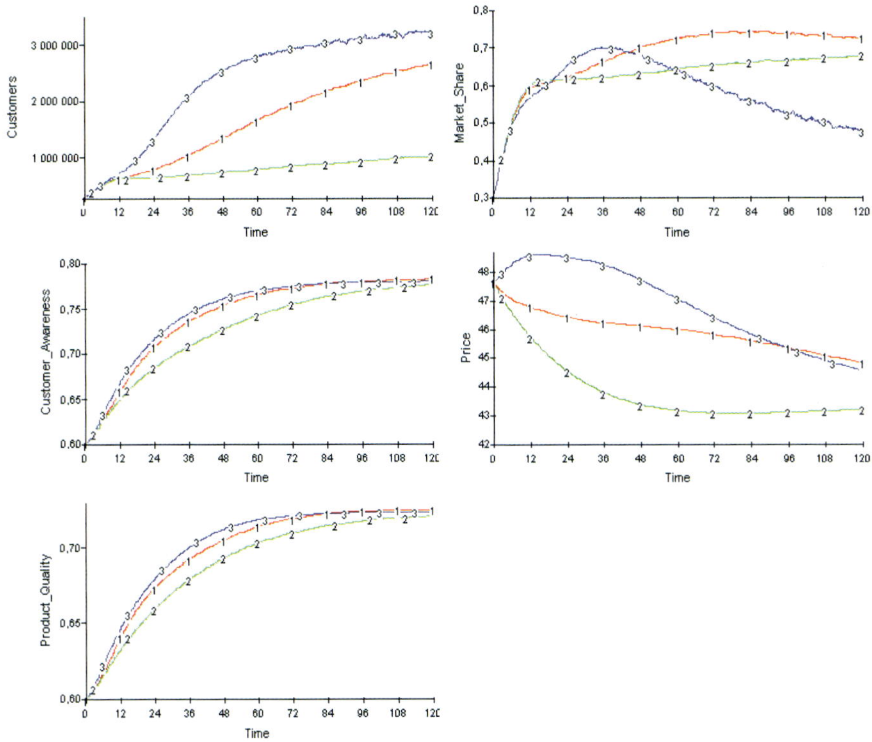

Figure 25: Results of important indicators for policy evaluation.

high price season for the Policy 3, the number of **Customers** is the highest too among the others. This means it is the highest revenue case.

In terms of **Market Share**, Policy 3 is still good until around time 44 at which Policy 1 passes by. However, one should not forget the **Potential Market** is enlarged by the effect of the company's good products and quality and awareness of the **Customers**. If you make your Customers aware of the industry, this would affect the entire industry so it grows too. As it grows your **Marketing Share** decreases by a certain amount. Therefore, it seems that though the **Marketing Share** is on the fall, one should pay attention that **Capacity** constraints are beyond the scope of this project. In fact, this is a good sign that model works well. It gives one clue about the future improvements and research about the project. If the company increased the **Capacity**, then there may be increasing or constant level of **Market Share** (Fig. 26).

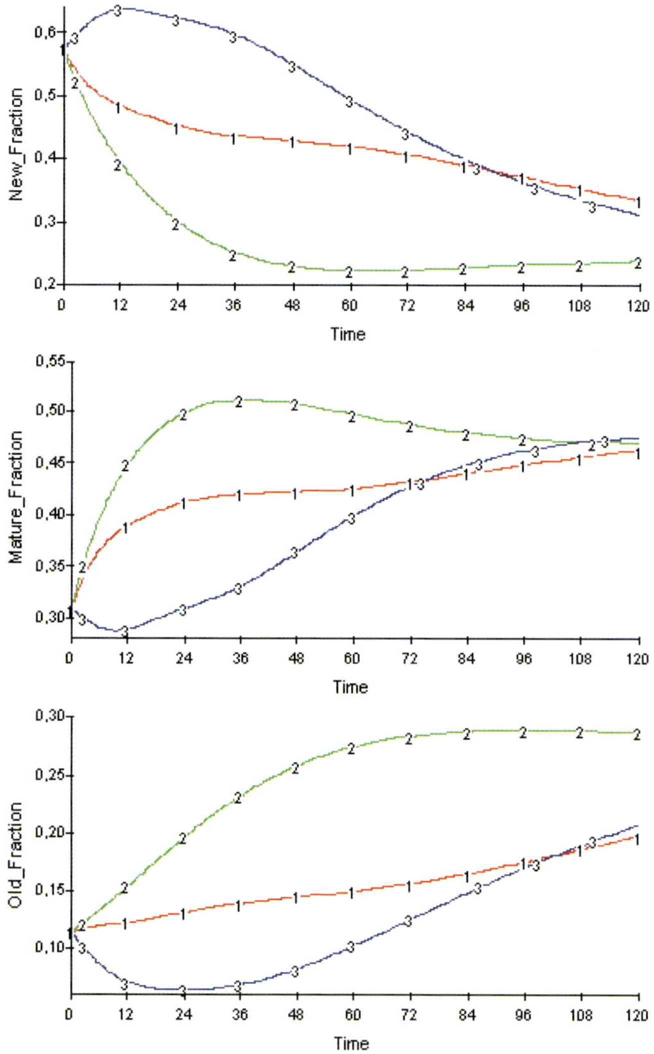

Figure 26: Results of product fractions for policy evaluation.

In terms of Product ages, one can see that, with Policy 3, **New Fraction** is almost always the highest. Policy 1 passes it by just in the end of the simulation. Policy 2 which was low R&D case shows that **New Fraction** is the lowest among all. **Old Fraction** is the lowest with Policy 3 almost always which is a good feature for a company, especially for an electronics company. Policy 1 becomes better in the end.

From all these points made, Policy 3 seems to be the best policy to be chosen among the alternative ones. Although there are some trade-offs, the manager can easily evaluate that Policy 3 is the preferred policy for the sake of the company. However, let's not forget that Policy 3 is a bit more expensive policy than the others. This can also be another future analysis for the project.

Policy Improvement: Policy 4 (Hybrid Policy)

Policy 3 is chosen to be implemented but a few improvement actions can be taken. Policy 4 is created for this reason with better and more feasible figures. Here is how Policy 4 is formed:

Policy 4: 1. Follow Policy 3 between time 0 and 60.
2. Shift to Policy 1 at time 60 and stay with it until time 90.
3. Shift back to Policy 3 at time 90 stay with it until the end.

The results can be seen below. **Curve 4** is Policy 4 (Fig. 27)

As one can see, Policy 4 is the replication of Policy 3 till time 60. The difference comes just after time 60. Between the time range 60 and 90, the **Market Share** is pretty much kept constant because of the shift to Policy 1 which has more marketing expenditure. For this period, **Quality** drops to a level but **Awareness** does not. The reason is that **Awareness** is fed with the **Marketing Budget** during which it is more than Policy 3. Price also drops significantly. The **Customer** base does not

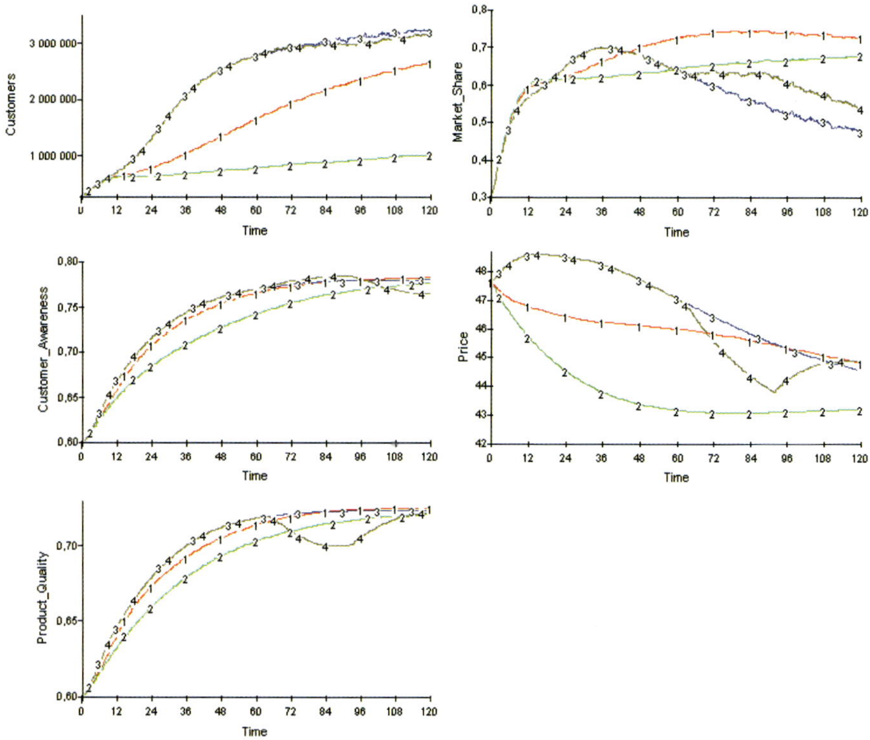

Figure 27: Results of important indicators for policy improvement.

change much. It is almost the same. After time 90, **Market Share** starts to drop again due to the fact that **Marketing Budget** is now limited to some degree. This time **Quality** starts to get up again due to high **R&D Budget**. **Awarenes**s starts to drop due to low **Marketing** again. **Price** on the other hand starts to increase.

At this point, it would be beneficial to make a comment about the effects of **R&D** and **Marketing** to the industry demand. From the analysis until now, one can say that **R&D** affects the industry demand more than the **Marketing** does (Fig. 28).

Policy 4 until time 60 is the same with Policy 3 which was mentioned before. When the shift occurs, the **New Fraction**

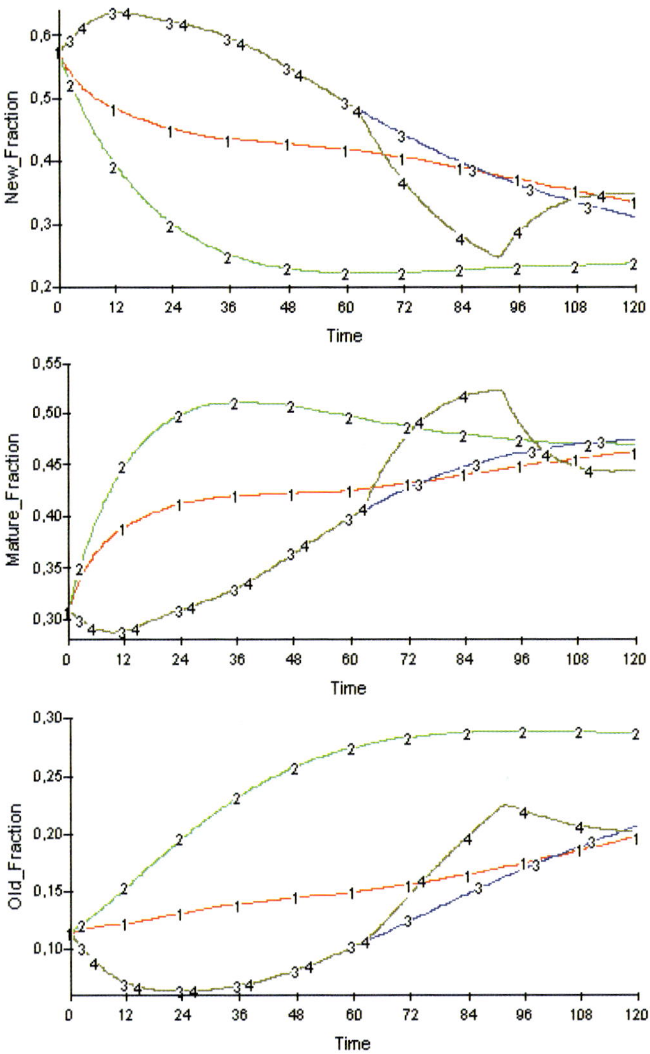

Figure 28: Results of product fractions for policy improvement.

starts to drop, but after the second shift to Policy 3 again, **the New Fraction** reaches a level that is higher than Policy 3 would do itself. Besides, the **Old Fraction** starts to decrease after the second shift.

Policy 4 seems to be a good hybrid of Policy 3 and Policy 1 which is also much cheaper than Policy 3 itself. The manager can choose Policy 4 for a less costly policy and if he wants to have a longer period of stable company. What can be done more is that he can play with these policies more by shifting from Policy 3 to policy 1 from time to time.

Conclusions

Key Findings

After the analysis of the policies, one sees that Policy 4 is a good policy for a company. Satisfying both the sales and revenues with the marketing share and customer base would be possible.

The reason why Policy 4 is the best among the studied alternatives is that many managers are evaluated based on their performance on market share. Even though they have good sales, there is always a question like why the market share is dropping. The current model enables the manager to control this issue, too. It is recommended for the manager to increase the **R&D Fraction** and decrease **Marketing Fraction**. By doing so, he or she can assure his or her position in the company and provide sales to the company. The policy may seem obvious, even intuitive, but it has a reason behind it. Most of the managers may have already been doing this, but are they really on purpose or is it just the market drive?

In alternative run (altrun), the model is run under the oscillating level of GDP Growth instead of a constant growth. It is recommended this time to increase the marketing when GDP is a downturn situation and increase the R&D share when GDP is on the rise due to the fact that when **GDP Growth** goes up, it is assumed that people get more purchasing power and they can afford more new products. If GDP is down, then the company should make marketing on the existing products more.

By investing in R&D and Marketing companies can enhance market growth by increasing the awareness among the customers so that they decide to buy the company's products which may result in increased market share.

Limitations and Suggestions for Future Research

The general framework of the model includes decisions such as marketing and R&D (for both the average quality of the products and aging new product launches). Production, inventory, backlog and issues such as delivery delay, competitors' reactions, and licensing are not considered.

This model may be enlarged by adding some variables such as Production, Inventory, Backlog, Capacity, Licensing, and some sectors such as Finance and Employees. By this way, it would be a more thorough and complete analysis. The analysis may be carried forward especially in terms of cost and profit sides so that the decisions can be made in a more reliable and complementary way.

Rivalry is another issue that may be analyzed in future studies. The company can identify certain types of rivals and group them into classifications so that the analysis gets easier as Warren (2002) and Scherer (1967) state. The company can have a group of rivals sectors added in the model.

Managerial Implications

Before attempting to solve resource allocation problems, it would be very helpful to understand the underlying structure of the problem at hand. SD is a useful tool dealing with causalities and nonlinearities, which are neglected, ignored or even feared by the most of static analysts.

References

Anderson, D. K., and Reynolds, S. E. (1997). Demand articulation in Japanese research and development. *Journal of Undergraduate Research,* **8** (1).

Baker, N., and Freeland, J. (1975). Recent advances in R&D benefit measurement and project selection methods. *Management Science,* **21**(10): Application Series, 1164–1175.

Barlas, Y. (1996). Formal aspects of model validity and validation in system dynamics. *System Dynamics Review,* **12**: 183–210.

Confessore, G. and Mancuso, P. (2002). A dynamic model of R&D competition. *Research in Economics,* **56**: 365–380.

Dickinson, M. W. (1999). Technology portfolio management: Optimising interdependent projects over multiple time periods. Master thesis, Massachusetts Institute of Technology (MIT), Boston, USA.

Ford, D. N., and Sterman, J. D. (1998). Dynamic modelling of product development process. *System Dynamics Review,* **14**: 31–68.

Garcia, R., Calantone, R., and Levine, R. (2003). The role of knowledge in resource allocation to exploration versus exploitation in technologically oriented organizations. *Decision Sciences,* **34**(2): 323–349.

Gill, G. K. (1992). Allegheny Ludlum: Research and engineering resource allocation, *Harvard Business School Case # 9–692–027,* 1–29.

Graham, A. K., and Ariza, C. A. (2003). Dynamic, hard and strategic questions: Using optimisation to answer a marketing resource allocation question. *System Dynamics Review,* **19**: 27–46.

Hauser, J. R. (1997). Research, development and engineering metrics. Accessed November 7, 2003, available at http://web.mit.edu/hauser/www/Pages/rdemetrics.pdf

Jenkins, W. (2005). Competing in times of evolution and revolution — An essay on long-term firm survival. *Management Decision,* **43**(1): 26–37.

Kameoka, A., and Takayanagi, S. (1999). A corporate technology stock model — Determining total R&D expenditure and effective investment pattern. *Proceedings of Portland International Conference on Management of Engineering and Technology.*

Lee, T. K. (1985). The joint decisions of R&D and technology adoption, *Management Science,* **31**(8): 959–969.

Leenders, M. A. A. M., and Wierenga, B. (2002). How the impact of integration of marketing and R&D differs depending on a firm's resources and its strategic scope. *ERIM Report Series Research in Management 68-MKT.*

Lyneis, J. M. (2000). System dynamics for market forecasting and structural analysis. *System Dynamics Review,* **16**: 3–25.

OECD. (1996). Technology, *Productivity and Job Creation,* **2**(2): 61.

Papadakis, M. (1995). The delicate task of linking industrial R&D to national competitiveness. *Technovation,* **15**: 569–583.

Poll-results. Accessed December 5, 2003, available at http://groups.yahoo.com/group/strateji/

Repenning, N. P. (2000). A dynamic model of resource allocation in multi-project research and development systems. *System Dynamics Review,* **16**: 173–212.

Rink D. R., Roden, D. M., and Fox, H. W. (1999). Financial management and planning with the product life cycle concept, *Business Horizons*, September–October, 65–72.

Scherer, F. M. (1967). Research and development resource allocation under rivalry. *The Quarterly Journal of Economics,* **81**(3): 359–394.

Shapiro, B. P. (2003). Bridge the gap between strategy and tactics with the magic matrix. *Harvard Business School Case # 9–999–008*, 1–24.

Souder, W. E. (1972). A scoring methodology for assessing the suitability of management science models. *Management Science,* **18**(10): Application Series, B526–B543.

Sterman, J. (2000). *Business dynamics: Systems thinking and modeling for a complex world*. New York, NY: McGraw-Hill Companies.

Trott, P. (2002). *Innovation management and new product development*. Financial Times + Prentice Hall, Harlow, England.

Wälde, K., and Woitek, U. (2004). R&D expenditure in G7 countries and the implications for endogenous fluctuations and growth. *Economics Letters,* **82**: 91–97.

Warren, K. (2002). *Competitive strategy dynamics*. London: John Wiley & Sons, Ltd.

Zoltners, A. A., and Sinha, P. (1980). Integer programming models for sales resource allocation. *Management Science,* **26**(3): 242–260.

CHAPTER 11

New Product Development: Performance Measurement Evaluation Matrix

Stefan Cedergren, Christer Norstrom* & Anders Wall†*

The ability to measure performance is both fundamental to and critical in successful product development. Previous research has focused on adding new measurements, not on evaluating those currently used. The performance measurement evaluation matrix (PMEX) is a tool to evaluate the performance measurement system based on success factors in the development of software-intensive products. Three case studies have been conducted to test the PMEX and the results indicate that the PMEX enables managers to determine explicitly what is and what is not measured. Time, cost, and quality seem to be the focus of the performance measurements while technology and planning activities are absent.

For a product development function to be successful, it must be possible to measure its performance (Davila *et al.*, 2013). Measuring performance is a classic and complicated research problem. This chapter stresses the importance of evaluating the performance measurements currently used within

*School of Innovation, Design and Engineering
Malardalen University, P. O. Box 883
Västerås, SE-72123, Sweden
†ABB Corporate Research, Forskargrand 6
Västerås, SE-72178, Sweden

the product development function from a management perspective, that of the research and development (R&D) director, product development portfolio owner, or others responsible for the product development function. In studying performance measurements from the point of view of the product development manager, a holistic system perspective of the product development function is adopted.

A vast amount of research within the area of performance measurements is available (Neely, 2007), mainly focusing on the design and the implementation of new performance measurement systems. However, the development and implementation of a new performance measurement system is a time-consuming and costly process. Few research studies focus on evaluating the performance measurement systems in current use. To address this issue, a tool for holistically evaluating performance measurements within this context is proposed in this chapter. Increased complexity stresses the need for models that can be used by teams to develop a shared understanding (Katz and Kahn, 1978). The real leverage in most management situations lies in understanding dynamic complexity, not detail complexity (Senge, 1990). Moreover, managers within product development are often aware that they are using a less than optimal measurement system but are unable to pinpoint explicitly its advantages and disadvantages. Add to this the problem of keeping measurements relevant to today's radically changing business and organizational context; "old" measurements are often not discarded and new measurements merely add to the confusion (Paranjape *et al.*, 2006).

In this research project, organizations developing industrial software-intensive products, in a business-to-business setting, are the unit of analysis. An industrial software-intensive product is usually a system consisting of simple, interconnected parts, usually software, electronic and mechanical components that in combination, exhibit a high degree of complexity from which emerges higher order behavior. The system is often long-lived and has been evolved by generations of engineers in large organizations, no single person understanding the entire system. These products usually have long life cycles and are becoming increasingly software-intensive. Product development activities are best described as incremental or evolutionary rather than radical in character. Normally, a platform or architecture is used as a basis, in order to manage the technical complexity of the product and shorten the development time and effort required for a new product. Examples of such products are found within industrial automation, commercial vehicles, telecommunication systems, and transportation systems.

Methodology

Complex research contexts often require a methodology which consists of various techniques rather than a methodology relying on a single method. To deal with the complexity of evaluating performance measurements within industrial product development, an approach consisting of a literature review, focused group interviews (Patton, 2002) and exploratory case studies (Yin, 2013) have been adopted. The case studies mainly involved single semistructured interviews (Robson, 2002) and analyses of the documentation of the performance measurement system concerned. This approach has been complemented with seminars and discussions of the results in both academia and industry. By adopting several data sources, triangulations of the findings are possible. An outline of the methodology including the major feedback loops is presented in Fig. 1.

This research began with a review of the relevant literature in order to identify the research gap and position this research in the existing literature. On the basis of the scarce focus on

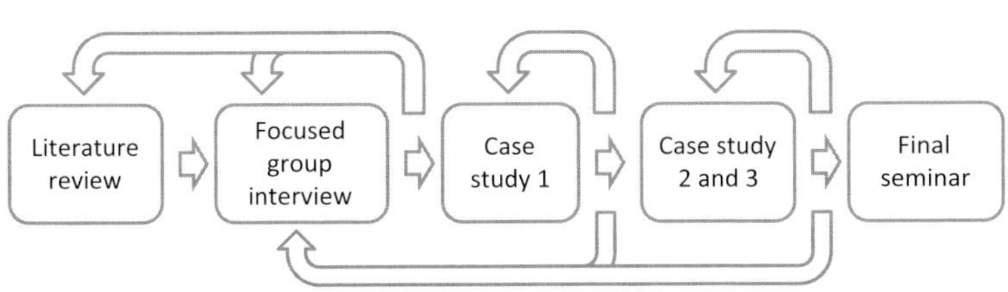

Figure 1: Outline of the research approach.

evaluating the evaluation system, within the domain of industrial software-intensive products in particular, an inductive research approach was decided on. The first step in developing an effective performance measurement system is to obtain relevant performance criteria and then define the required measurements accordingly (Gharajedaghi, 2006). A review of the success factors identified in the literature dealing with product development was therefore conducted. Since the focus of this research is on the domain of industrial software-intensive product development, and most studies of success factors relate to more general domains, it was decided to use focused group interviews in order to determine the factors managers identify as important for success.

The focused group interview was performed with senior managers of product development from eight companies. The managers participating were selected on the basis of their experience in developing industrial products. They represented international companies, with the exception of one management consultancy firm, all being active in Sweden. They all have extensive experience in developing industrial software-intensive products within transportation, telecommunications, automation, and the automotive industry. The participating companies are not competing in each other's markets and have different products in their portfolio. However, they share similar difficulties and challenges in achieving a high-performance product development function and being able to measure its performance. The focused group interview was carefully planned and began with a presentation of an initial differentiation of the product development function into planning and implementation activities. This was followed by dividing the participants into smaller groups to discuss, based on their experience, what is important for success in each of the two categories. Each group presented and discussed their results in the larger group, resulting in a list of important factors. This list of success factors was then analyzed and categorized by the authors. The resulting initial framework of success factors was then presented and discussed at a seminar with the participants in the focused group interviews. A comparison of the results with the findings in the literature was also performed. An overview of the different feedback loops is shown in Fig. 1.

The initial framework of success factors, that is, performance criteria developed from the focused group interview was then used as input to the exploratory case studies intended to determine how performance measurements are used in practise. The explorative multiple case studies were performed in accordance with the approach presented by Yin (2013). A case study research strategy focuses on understanding the dynamics present within a single setting (Eisenhardt 1989) and may therefore be suitable for investigating the use of performance measurements within industrial product development. The aim of the exploratory case studies was to obtain a deeper understanding of the planning and implementation processes within product development in particular.

The three case companies were selected from the companies participating in the focused group interview, all being divisions within Fortune 500 companies with a high R&D level in relation to revenue. The three organizations develop industrial products in the fields of industrial automation and heavy vehicles. A total of 41 semistructured interviews were held at the three case companies to identify how they perceive performance and how it is measured at different levels of the organizations. The first case study started with an initial set of selected respondents, together with the senior manager representing the company in the focused group interview. During these interviews additional respondents were identified and interviewed as necessary.

Based on the first case study, time, cost, and quality were identified as common areas of measurement. It was therefore decided to extend the initial framework of success factors to include these categories even though they are not to be considered success factors but rather as the result of succeeding with the success factors. The second and third case studies did not require any changes in the framework, but added to knowledge of how performance measurements were used. The findings from each case study were presented at seminars at each of the three companies.

Limitations

In order to ensure the reliability of this research, a far from linear and straightforward research approach has been adopted, as illustrated in Fig. 1. Focused group interviews have several advantages (Patton, 2002). In this research, they were chosen because the interaction between participants enhances the data quality and because it is also relatively easy to evaluate the consistency of the data obtained. The climate of the focused group interview was constructive, the participants readily sharing and learning from each other's experiences. It is also a cost-effective way to gather empirical data (Robson, 2002). Moreover, focused group interviews tend to work well when the participants, though sharing similar backgrounds, are strangers to each other (Patton, 2002). Focused group interviews are often used in conjunction with other methods like individual interviews (Robson, 2002). Data collection is generally difficult in groups, and the findings were therefore recorded during the interviews in such a way that all participants could see the data collected and acknowledge its validity. Several iterations between various stakeholders have been made. Seminars and feedback to the respondents involved have been an important part of this research study. After conducting the exploratory case studies a final seminar with the participants from the focused group interviews was performed to conclude the study, to draw common conclusions from the case companies.

In common with all research studies based on qualitative data, this study has its limitations. With respect to generalizability, it is difficult to draw any conclusions outside the area of industrial software-intensive products. However, the main ideas behind this research study could easily be applied within other domains in order to identify the relevant performance criteria to thereby evaluate the performance measurements used in the particular context. Moreover, the eight companies represented in the focused group interview all develop different products and are active in different markets but still share a common view of the success factors. Thus, the results from this study should be applicable beyond the domain of industrial software-intensive products.

Related Work within Product Development and Performance Measurement

Transforming an innovative idea into a successful product is by defining a unique procedure and as such cannot be treated as a purely bureaucratic process. It must be treated as a separate project, a finite activity with its own objectives and resources, and above all, its own leadership (Goffin and Mitchell, 2005). Measuring product development performance over time within such a project becomes complex due to the resulting inherent uncertainty (Bremser and Barsky, 2004). The area of performance measurement has interested researchers with different functional backgrounds and there is a vast amount of research available. Few studies have, however, analyzed the product development process from the performance measurement system perspective (Jiménez-Zarco *et al.*, 2006). The Stage-Gate model (Cooper, 1990) establishes the main guidelines for developing such an analysis. This model presents the product development process as a complex system consisting of two independent and parallel processes: the product development process itself and the evaluation process.

The Product Development Process

The term product development is often used without a proper definition. The research presented in this chapter emphasizes a holistic view of product development by proposing the following definition:

> "Product development is the set of activities beginning with the processes and tools used to perceive a market opportunity and ending in the production, sale, and delivery of a product fulfilling that market opportunity".

This definition is an extension of the popular definition provided by Ulrich and Eppinger (2008), by emphasising the importance of the processes and tools used in the early stages of the process when the scope of the development activities is determined. In this research, the product development function is to be considered successful if its products not only satisfy the needs and requirements of customers, but also generate profits to its shareholders, and creates value to its stakeholder at large.

The process of developing new products within industry is diverse both in the sense of novelty and the type of products developed. The reputable Stage-Gate model (Cooper, 1990), based on Booz, Allen, and Hamilton's model (BAH), is both a conceptual and an operational model intended to move a new product from the idea stage through to market launch and beyond (Jiménez-Zarco *et al.*, 2006). The generic Stage-Gate process includes stages and gates (Cooper, 2008). There are close similarities between the generic phases in the product development process and the stages in the Stage-Gate model. During the stages in which the development work is performed, the project team completes activities to reduce business risks and advance the project to the next gate at which the Go/Kill and prioritization decisions are made (Cooper, 2008). The focus during the gate decision procedures should be on three key issues: quality of execution, business rationale, and the quality of the action plan (Cooper, 2001).

Studies show that 60% of enterprises use a Stage-Gate or similar process during product development, whereas 39% make no use of formal development processes (Griffin, 1997).

Further, results indicate that companies modifying their formal product development process improve their efficiency while not significantly sacrificing product novelty (Ettlie and Elsenbach, 2007). The Stage-Gate model usually involves four to eight stages and gates in a process modified by the company (Jiménez-Zarco *et al.*, 2006).

Critical Success Factors in Product Development

Research literature identifies, and industry best practices confirm, a vast number of success factors that contribute to successful product development (Tang *et al.*, 2005). The research area of critical success factors has matured. In the most recent meta study of critical success factors in product development previous findings are confirmed (Evanschitzky *et al.*, 2012). Still, no prescribed common criterion can explain how successful new products are created (Poolton and Barclay, 1998). The SAPPHO project (Rothwell *et al.*, 1974) was the first effort to analytically compare commercially successful and unsuccessful products in the same market. The conclusions from this project are that successful companies have a much better understanding of customer needs, attend more to marketing and advertising, perform product development more effectively, encourage more use of outside expertise, and authorize and promote responsible and experienced professional employees to senior management levels. Their conclusions were that professional employees and good management skills are the key to success (Abdel-Kader and Lin, 2009). The concept of a limited number of factors that directly affect the outcome and underlie successful performance of the product development process is appealing to every manager. As a result, there is a vast amount of research available within the area of success factors within product development (Ernst, 2002). Normally, success factors are identified either at the business unit level or at the product level by comparing a successful business unit or product with one less successful.

Tang *et al.* (2005) identified a distinct set of success factors for product development that are statistically accurate predictors of the outcomes of profit, market share, customer satisfaction, organizational effectiveness, and product quality, specific to the product. These include the following: *Leadership*, *Organizational culture*, *Human resources*, *Information*, *Product strategy*, *Project execution*, *Product delivery*, and *Result*. As a success factor, leadership involves the personal characteristics of the project leader, delegation of authority, and whether there is clear strategic direction of the project. Organizational culture determines the extent to which the management takes advantage of the established values of the employees to improve project outcomes. Human resources are a success factor requiring management action to improve skills and the working environment. Information is concerned with the treatment of information as a valuable asset, its quality, and whether it is systematically collected, shared, analyzed. Product strategy includes the product planning processes and the extent to which they promote readiness for product development and product delivery. Project execution involves the key issues of the product development process. Product delivery considers to what extent manufacturing, sales, service, and support are considered; or whether the product is just "tossed over the wall". Results evaluate the project from different directions such as financial and market, customer satisfaction and loyalty, organizational effectiveness, product results, and benchmarking.

In a thorough review of critical success factors by Ernst (2002), the following categorization, as previously developed by Cooper and Kleinschmidt (2007), was adopted: *Customer integration*, *Organization*, *Culture*, *Role and commitment of senior management*, and *Strategy*. In another review by Adams *et al.* (2006) drawing on a wide body of product innovation literature, the following seven categories are identified as important in the product innovation process: *Input management*, *Knowledge management*, *Innovation strategy*, *Organizational culture and structure*, *Portfolio management*, *Project management*, and *Commercialization*. Bessant and Tidd (2007) argue for the following success factors in product development: *Market knowledge*, *Clear product definition*, *Product advantage*, *Project organization*, *Top management support*, *Risk assessment*, *Proficiency in execution*, and *Project resources*. Product advantage involves product superiority in the eyes of the customer, for example, unique benefits to the user at a high performance-to-cost ratio. Market knowledge, that is, assessment and understanding of customer and user needs is critical. A clear product definition by defining target markets, clear concept definition, and benefits to be delivered is necessary before the development begins. Holistic risk assessment based on information from marketing, technological, manufacturing, and design sources must be incorporated in the business and feasibility studies. The use of cross-functional, multidisciplinary teams carrying responsibilities is important within the project organization from the beginning to the end. Project resources include financial, human skill, and material resources. The company must possess the right skills to manage and develop the new product. Proficiency in execution includes all the activities of the product development process. Top management support is important through the complete product development process from concept to launch.

Performance Measurement

The field of performance measurement has inspired numerous researchers with functional backgrounds as varied as accounting, operations management, marketing, finance, economics, psychology, and sociology, all actively working in the field (Neely, 2007). Although measurement is both fundamental and critical to success in product development (Davila *et al.*, 2013), performance measurement in product development is still nascent (Tatikonda, 2008). There are several reasons why it is important to measure performance. Neely (2007) propose four different reasons for measuring performance: check position, communicate position, confirm priorities, and compel progress. This illustrates an important fact: measurement alone is not a direct value-adding activity; value is achieved when the result of the performance measurement system is actively and adequately used by management.

The most widespread and cited performance measurement system is the Balanced Scorecard (Neely, 2005), introduced by Kaplan and Norton (1992). Financial measures alone cannot adequately reflect factors such as quality, customer satisfaction, and employee motivation (Driva *et al.*, 2001). The reason behind the development of the Balanced Scorecard was to balance the financial perspective with the perspective of customers, learning and growth, and internal business processes. Implementations of the Balanced Scorecard are not always successful and translating the Balanced Scorecard into concrete action remains a problematic area. Bremser and Barsky (2004) outline a framework to integrate the Balanced Scorecard in R&D management. The Balanced Scorecard has, however, not had the same success within product development as it has within more general business performance measurements. Balanced measurements are intended to provide a balance by including measurements of success, external as well as internal, together with measurements intended to give an early indication of future business performance as well as a record of what has been achieved in the past (Bourne *et al.*, 2000). Leading indicators include measures affecting the process, while lagging indicators measure the results of processes already performed. There are several other different classifications of performance measurements. Two basic distinctions are between quantitative and qualitative measurements. For example, computational methods clearly lead to a quantitative value, for example, time to market has been six months, whereas assessment methods usually result in a qualitative indication of the metric value, for example, time to market has been "good" or "unsatisfactory" (Kerssens-van Drongelen *et al.*, 2000).

Research within performance measurement is often focused on the design and implementation of a performance measurement system. Several models and frameworks other than the Balanced Scorecard are available but few of them focus explicitly on the product development process. A newly developed framework is the Performance Prism (Neely *et al.*, 2001) that emphasizes a more holistic approach to the stakeholder perspective of performance measurements than the Balanced Scorecard. Moreover, few studies within performance measurement involve a holistic evaluation of the currently used performance measurement systems. Merely half of the 10 performance measurement systems, identified in (Pun and White, 2005), have any kind of evaluation of the current performance measurement system in their process of designing the new system.

Performance measurements are important as an aid in determining the priorities to be given different activities, and as a means of guiding teams by indicating how they are performing and where improvements would be most beneficial. However, performance measurement must be kept in perspective; it must support the product development process and goal attainment (Nixon, 1997) based on the business strategy. Linking pursued strategy to a few explicit measures is a fundamental rule within the performance measurements literature (Davila *et al.*, 2013). This implies the importance of continuously evaluating the performance measurement system. Process management theory suggests that one should not only implement the correct processes, one should also monitor how well the process is operating and, if necessary, intervene in a timely manner (Syamil *et al.*, 2004).

The Performance Measurement Evaluation Matrix

This section outlines the proposed PMEX. The PMEX provides product development managers with a tool to evaluate the performance measurement system they are currently using.

Without a structured method, such as the one proposed, it is difficult to assess what is measured and, perhaps more important, assess what is not measured by the current performance measurement system. Product development is a dynamic and diverse task involving many different competences and functions within a company. As a result, the process of product development differs between companies and between domains within industries. This research focuses on evaluating the performance measurements used by companies developing industrial products. It was therefore begun by identifying important success factors within the development of industrial products. The first result of this research, a framework for performance evaluation in innovation is presented in (Cedergren *et al.*, 2010). The product development literature includes several studies identifying success factors but such studies relating to the explicit area of industrial product development are absent or their results have not reached wide acceptance. Conceptually, we have divided the product development process into three different categories based on our findings: *Planning*, *Implementation*, and *Sales and Delivery* (see Fig. 2).

This categorization is in line with the definition of product development previously proposed. The *Planning* activities typically involve decisions regarding what product to develop, especially the planning and concept development of the generic phases of the product development process. The *Implementation* activities are more operational in designing and constructing a product, typically involving system level

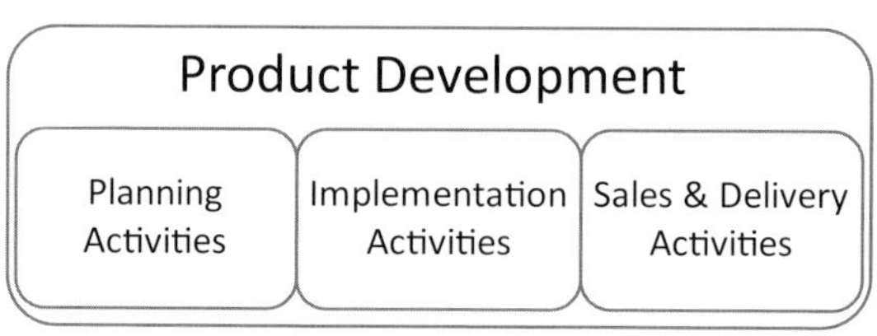

Figure 2: Proposed categorization of the product development process.

design, detail design, test and refinement, and the production ramp-up. The final part, *Sales and Delivery* activities are important in ensuring the completeness of the product development process and its overall success. All the three parts of the product development process require unique specific competences and objectives. The product development process cannot be considered successful until the targeted customer needs are fulfilled and the new product generates profit. Moreover, it is important to recognize the different competences needed in the *Planning* and *Implementation* activities of the product development process. From a performance measurement perspective, it is vital to differentiate between *Planning* and *Implementation*, since their objectives differ. The *Sales and delivery* activities will not be further developed in this chapter.

When a performance measurement system is to be evaluated, it is important to address *what* is to be measured and *when* it is to be measured. The motivation for a particular measurement, the *why*, is also a central issue in the process of evaluating a performance measurement system, especially since every measurement involves a cost. To address *what* is to be measured, the success factors in industrial product development are to be identified by means of focused group interviews. The results of these focused group interviews are then categorized and the result is presented according to the product development planning and implementation in the following.

Product Development: Planning

The first part of the product development process, and certainly the most important, is to decide why the product should be developed. It is during the product development planning that the upper boundary for overall success and profitability of the product development is set. The overall objective of the product development planning is to determine customer needs and transform them into a product that utilizes the company's resources in the optimum way thereby maximizing future cash flow. Two main questions to be answered during product development planning emerged during this study. The first question is *what* to develop and *why*, the second question is *how* and *when* to develop it. *What* and *why* are vital questions, since they set the boundaries for both the technical specification and the future cash flow. In a value-creation perspective, once a company has decided *why* a product should be developed and *what* that product should be, the future value of the product development investment is limited since the technical solution and targeted market are decided upon. This is an important aspect of the product development process that needs to be managed to make optimum use of the available company resources (Fig. 3).

The how and when questions in the product development planning focus on making optimum use of the company's resources with efficient project execution as an important objective. A key success factor for how and when is not to attempt to start product development implementation activities if the required resources are not immediately available. It is common for companies to initiate project after project without securing the key competence (Seider, 2006). If a new product development project is initiated in an already fully utilized organization, it will only slow all projects down. Technology planning that supports and speeds up product implementation activities is a vital success factor (Wheelright and Clark, 1992).

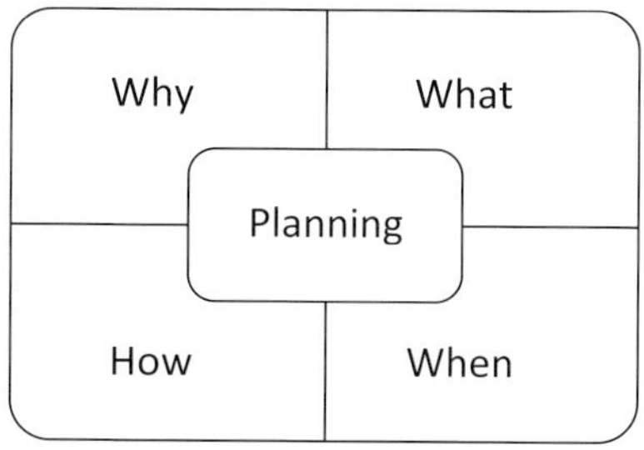

Figure 3: Product development planning activities involves: why, what, how, and when to develop a product.

The planning ends and the implementation begin when the firm decides to execute the product development project, that is, implements the planned product. Table 1 illustrates the success factors that product development managers in our studies consider most important in the product development planning activities.

Product Development: Implementation

The product development implementation activities consist of project execution that is, the realization of that which was specified in the product development planning, as efficiently and effectively as possible. The ultimate success of the product development implementation is to deliver exactly that which was specified within the budget, on time, and with the specified quality. In the implementation activities, there are several categories of factors affecting the performance, shown in Fig. 4. These have been separated into four main categories influencing the success of the product creation according to the analysis of the interviews and workshops: *management*, *process*, *technology*, and *people*.

The implementation activities of the product development process could be viewed as a manufacturing process, since the best possible outcome is to deliver exactly what was specified in the product development planning. However, for this to be a reality, it is essential that the technology supports the project with predevelopment and reuse. Furthermore, it is important to have a properly planned project for example, a front-loaded project in order to achieve an efficient and predictable implementation process. It is also vital for the product development implementation to provide the people involved in the project with an understanding of what is needed of them. The ultimate success of product development implementation is to achieve the required time-to-market with sufficient quality. In order for management to make optimum use of the available resources, it is important that the project members find their assignments

Table 1: The important success factors in product development planning.

Why and What		How and When	
Market environment analysis	Involves different aspects: technology, competitors, the customers' future business and processes, market knowledge, and so forth.	Technology roadmap	Develop the technology needed to support the product roadmaps.
Customer needs and wants	The ability to fully understand the customer needs and wants.	Metrics	Different metrics assisting the decision making.
Business case	Clearly specify what kind of profit this product generates and why.	Organization	It should have clear responsibility, mandate, culture, competence, and roles to support the planning.
Product roadmaps	A clear plan of how the product will evolve in the future.	Ownership from top management	It is important that the CEO understands how the product development process will generate future revenues and profit.
Risk management	The ability to assess risks and work active with them.	Planning competence	Understanding all the aspects: technical, market, economic, production, purchase, and so forth needs and address them.

Source: Cedergren *et al.* (2010).

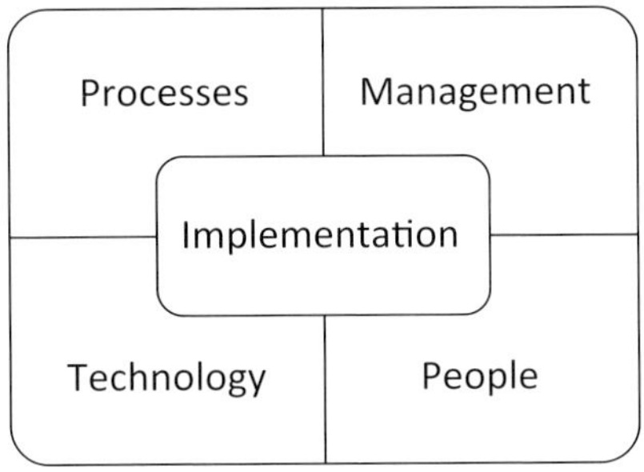

Figure 4: Product development implementation involves: processes, management, people, and technology.

professionally challenging, leading to accomplishment, recognition, and professional growth (Kahn, 2005). A study presented by Koehler and Weissbarth (2004) reveals that most new products, from automobiles to washing machines, are overengineered because the customer needs are not communicated and managed properly. In the implementation phase, it is important for management to continuously update and communicate organizational goals and project objectives to the project members. Management should also communicate the relationship and contribution of individual activities to the overall product development process and business case (Kahn, 2005). Tables 2 and 3 illustrate the results from workshops and interviews with respect to success factors in product implementation activities.

Comparison With the Success Factors in Literature

It is interesting to compare success factors identified within the domain of developing industrial software-intensive products with those presented in the relevant literature. Table 4 illustrates how the categorization of success factors identified in this research and presented in Tables 1–3 map to the success factors identified in the literature presented earlier.

Table 2: Success factors for the product development implementation.

Processes		People	
Process quality	The maturity of the processes	Feedback	Feedback to the people involved in the project to further develop their competence.
Clear development process	In the sense that everyone in the organization understands and is able to execute the processes in use.	Culture/ attitude	In the global world of today it is important to have all involved working together as a team.
Tools	Updated tools that support the product development work in the best way possible.	Organization	Important that the organization evolves with the changes that occur in the firm and thereby support projects the best way possible.
Industrial structure	Meaning that the right support systems are in place and can be used by the projects.	Resources	Important to have motivated and the required amount of resources available for the project.
Clear metrics	The use of metrics will improve the understanding of the performance of the process.	Competence	Involves securing a diverse and excellent competence in the company
Requirement management	A structured way of handling requirements.	Incentives	Could be in the form of bonuses and other rewards.

Source: Cedergren *et al*. (2010).

Table 3: Success factors for the product development implementation.

Management		Technology	
Professional project implementation	Important to have skilled project leaders for effective project execution.	Technical platform/ architecture	Makes it possible to share technology and thereby cost between projects/ applications
Multi project/ portfolio management	The company must be able to handle multiple projects and maintain effective project execution.	Predevelopment of technology	Shall support the implementation to improve time to market and quality
Risk management	All risks must be identified and assessed.		
Handle dependencies	Dependencies could involve business, resources, technical issues, and project.		
Global and local development	Find the right setting for what should be developed where.		
Clear objectives/ requirements	Management must be clear about what is expected from the people involved in the project.		
Supplier/partners	The ability to handle suppliers and partners during the development.		

Source: Cedergren *et al*. (2010).

Table 4: The success factors identified in this research mapped to the success factors identified in the literature.

	Tang *et al*. (2005)	Ernst (2002)/ Cooper and Kleinschmidt (2007)	Adams *et al*. (2006)	Bessant and Tidd (2007)
What	Product strategy	Customer integration, strategy	Portfolio management Innovation strategy Commercialization	Market knowledge clear product definition, Product advantage
Why	Product strategy	Strategy	Portfolio management Innovation strategy Commercialization	Market knowledge, product advantage clear product definition
How	Product strategy	Strategy		Market knowledge clear product definition
When	Product strategy	Strategy	Portfolio management Commercialization	Market knowledge

Technology				
Management	Leadership, organizational culture, Information, human resources	Organization, role and commitment of senior management, culture	Innovation strategy, Knowledge management Project management Organizational structure	Project organization, top management support
Process	Project execution, Information, product delivery results	Product development process	Input management, knowledge management, project management	Risk assessment, proficiency in execution
People	Information	Organization, culture	Organizational culture Input management	Project resources, proficiency in execution

It is difficult to compare directly success factors from the literature with those identified in this research because they are defined with different levels of abstraction. One example of this difficulty is strategy, identified by Ernst (2002) and Cooper and Kleinschmidt (2007) as an important success factor, which is mapped into all of the categorization *why*, *what*, *how*, and *when*. In the comparison of the success factors, the main objective is not to highlight the detailed variations; it is the major differences which are important. With this in mind, a first analysis of the mapping in Table 4 indicates that the technology category is not directly addressed by any of the other studies. This is surprising but might be explained by other studies focusing on a wider set of companies and products while this research explicitly focuses on the development of industrial software-intensive products. In the context of developing industrial products, technology is, per definition, an important aspect of product development performance. The technology aspect involves, for instance, platforms or product-line architectures that are used across a set of related products making it possible to share and reuse technology, thereby sharing cost between products and applications. Predevelopment of technology supporting the product development implementation is another factor effecting, for example, time to market and the quality of the developed product. Moreover, the technological infrastructure, for example, a systems' architecture can have both positive and negative effects on the product development performance. The architecture may exhibit different levels of inner-quality attributes such as evolvability, flexibility, testability which have effects on the performance when evolving a long-lived system. Christensen's (2003) well-known study within the disk drive industry highlights the importance of this issue. Both the rate of improvement of a technology's performance and the rate at which the technology is adopted in the market have repeatedly shown to conform to an s-shaped curve (Schilling, 2006). The ability to assess when the currently used technology reaches the end of such an s-curve, and hence is in need of, for example, improvement of the inner qualities of the architecture, would clearly be beneficial, especially for the efficiency aspect of product development performance.

The *how* and *when* questions in the product development planning are less emphasized in comparison with the questions of *what* and *why* in Table 4. The aspect *when* is important since it is a common phenomenon to overload the product development portfolio in the search for higher efficiency. However, such an overload often results in an increased product development project lead time. In a product development manager perspective, the success of the overall portfolio is of more importance than the performance of an individual product development project. Studies focusing on the success of an individual product development project are likely to miss out on the importance of the *when* perspective of the product development portfolio performance. Companies must carefully balance new initiatives against their implementation capacity to effectively execute their projects. Initiating new projects in constrained organizations can devastate overall department performance; few organizations actually measure development capacity or properly evaluate the impact of new initiatives on commitments already in progress (Seider, 2006). In the product development implementation, all the studies highlight *management* as the most important success factor.

The Performance Measurement Evaluation Matrix

In evaluating performance measurement within the product development process in a holistic perspective, why and what is measured is important but when it is measured is just as important. In this research, the Stage-Gate process is proposed as a means of addressing the when to measure what. Since the

Stage-Gate process represents the different phases of a product development project it is natural to also use it to represent the timeline of the sampled measures. The PMEX (Fig. 5) has the different phases of the Stage-Gate process as one dimension of the matrix and the categorization of important success factors for developing software-intensive products, as identified in this study, as the other dimension. The success factor categorization included in the PMEX represents what is important to manage in order to enable a high performing product development process. It may be tempting to design a performance

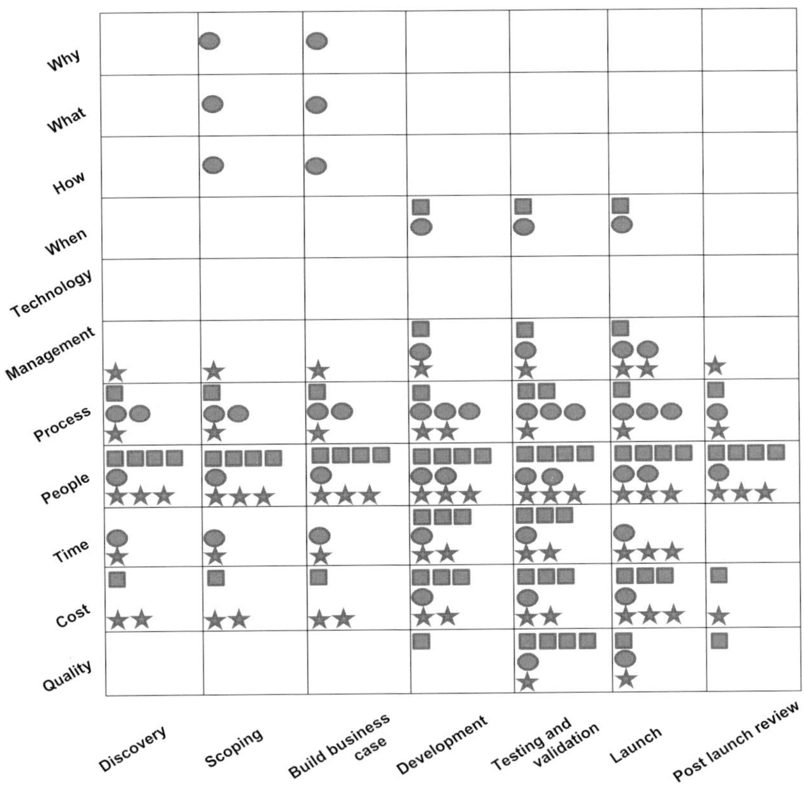

Figure 5: The PMEX applied to the three companies. Stars represent case company 1, circles case company 2, and squares represent case company 3.

measurement system that covers every square of the PMEX. This is not the intent of the PMEX and even if it is achieved, it probably would be difficult to make use of all the information. Instead, the PMEX should be viewed as a conceptual tool for product development managers in considering the performance measurement system, in the sense of what needs to be measured, why it should be measured, and when it should be measured.

Applying the PMEX in Practice

The best way to test the PMEX is to apply it in practice, this being done through three exploratory case studies within the domain of industrial software-intensive products. From the multiple case studies, different performance measurements of the three organizations were identified and mapped into the PMEX. The first result from case study 1 was to extend the PMEX to also include time, cost, and quality, not because they are success factors in the product development process, but because many measurements involve time, cost, and quality. It is surprising that time, cost, and quality were never mentioned in the focused group interviews when the success factors where elicited and analysed by the senior managers participating in our study. One interpretation can be that success factors can be related to leading indicators, while time, cost, and quality are lagging indicators of performance, enabled by the success factors. Hence, when the focus is on performance criteria, time, cost, and quality are not considered.

The proposed PMEX, with an indication of what is measured by the three case companies, is shown in Fig. 5. The performance measurements from the three case companies were successfully structured into the PMEX. The stars represent the measurements used by case company 1; the circles represent the measurements used by case company 2; the squares represent the measurements used by case company 3. It was decided to map the measures into the matrix in order to emphasize the presentation and usage of the PMEX as an evaluation method, rather than to focus on the specific metrics used by these three case companies.

It should be noted that a star, a circle, or a square does not represent a unique measure. It could be that the same measure is used in a number of squares of the PMEX. An example of this is that the circle in the *why*, *what*, and *how* lines, measured during the scoping and build business case stages, is the same measure. It is a measure of how the prescriptive process of early stages in the product development process is followed. Multiple entries of stars, circles or squares represent multiple measures applied in the same area. A further result of the analysis is that the three case companies seem to measure what is easy to measure and not what managers consider to be important. If the measure of fulfillment of the early stages of the product development process is disregarded, there are hardly any measures within the planning activities (*why*, *what*, *how*, and *when*) or the technology aspects of the product development process. This may be negative in the important effectiveness perspective of the product development process. It may be the result of companies focusing on the efficiency and not the effectiveness of product development. Another interesting finding is that none of the three case companies measure the aspects related to technology, despite the potentially large influence it may have on, in particular, product development efficiency. Since all the companies studied are successful, and have an interest in increasing their product development performance, it could be that this information is managed as tacit knowledge. Tacit knowledge is defined as knowledge that cannot be articulated or verbalized (Foos *et al.*, 2006); it is a knowledge that resides in an intuitive realm. Since the subject of tacit knowledge transfer, content, and process is poorly understood (Foos *et al.*, 2006) it may be a substantial risk to treat the technology and planning aspects of the product development process in this way. If that is the case, it may be difficult to manage the planning activities as a process and thereby enable continued improvement.

Experiences From the First Verification of The PMEX

The first indications from tests of the PMEX show promising results. The PMEX has been tested at three case companies and within these the PMEX has been received as an insightful and useful way of evaluating the current performance measurement system. In all the studied cases the PMEX clearly illustrate what is and what is not measured. The main idea of the PMEX is to holistically evaluate the performance measurement system and use it as a conceptual tool when performance measures are discussed. Moreover, all case companies have a clear potential to further develop their performance measurement system since the success factors identified by the participating companies are disregarded to a great degree, especially the planning activities. This finding is supported by the literature, for example, in a study by Hertenstein and Platt (2000) companies seldom reflect their strategies in their performance measurement system. Furthermore, the findings that technology is not explicitly perceived as an important success factor in the product development literature and at the same time is disregarded by the performance measurement system are both interesting. This is particularly so since most of the product development within companies developing industrial software-intensive products is incremental development of long-living systems rather than development of completely new products. Product-line architectures or platforms are often used and shared between products. It is our experience that the important inner quality of such architectures declines over time if quality is not actively managed by the R&D organization and the product development management. Reasons for this include poor communication of important architectural decisions and constructs leading to architectural disintegration when new features are implemented in ways that violate the rules set by the architecture, the introduction of new features incompatible with the current architecture, changes in business context not supported by the current architecture, turn-over of engineers resulting in the loss of knowledge. For a company developing industrial products it is therefore a clearly competitive advantage to have its technology evolve parallel with the business context, thereby supporting an efficient and effective implementation of new features and applications.

Conclusions and Future Research

In this chapter, the PMEX, a tool for evaluating the performance measurement system of a company's product development function from a product development manager's perspective, has been proposed. An effective performance measurement system is based on relevant performance criteria. Hence, the PMEX has the success factors within the development of industrial products as one dimension and the phases of the Stage-Gate process, representing the timeline, as the other dimension. One benefit of the PMEX is the possibility, for a product development manager, to holistically evaluate what is measured and maybe more importantly, what is not measured. The PMEX may also be used as a conceptual tool to reason about the performance measurement system, making it possible to initiate discussions of *what* is measured, but also *why*, and *when* something is measured. Furthermore, as the PMEX also illustrates what is not measured, it can be used when changes or new measurements are to be added, in order to ensure a performance measurement system that measures what is important in a product development manager's perspective.

Another conclusion based on the result of using the PMEX is that the technology aspect of the product development process is not measured by any of the three case companies. This is surprising since the case companies as well as the focused group interview acknowledge technology as an important success factor. Moreover, a study of the literature relating to different success factors within product development also indicates a disregard of the technology aspect of product development performance. The effect of not measuring the technology aspect is difficult to assess since the impact is mostly long term. Further research is needed, focusing on possible success factors and measures that can be used to address the technology aspects within the development of industrial products. Overall, it seems like an explicit link between what managers identify as important success factors and what is measured is weak. One interpretation is that success factors may be regarded as leading indicators, thus more difficult to quantify in measurements compared to lagging indicators such as time, cost, and quality.

A fundamental rule within the performance measurement literature is to link the strategy pursued by an organization with its measurement system (Davila *et al.*, 2013), but important strategic factors such as the product development planning and the technology aspects were not emphasized in existing performance measurement systems. Instead, an overall conclusion from the case studies is that there seems to be a tendency to measure something because it is possible to measure, rather than because it is important to measure. As a result, there are areas in the PMEX, see Fig. 5, which are covered by up to four different measurements. This is far from cost effective but by means of PMEX this phenomenon can be detected.

At least one of the participating companies will address the weaknesses of their performance measurement system in the early phases of the development process as a result of this study. The problem of how to design performance measurements that address these issues remains. Therefore, since the development of an effective performance measurement system iteratively should deal with both performance criteria and performance measurements (Gharajedaghi, 2006), future research needs to focus on developing performance criteria relevant to technology and planning activities for use in the design of performance measurements in these areas.

References

Abdel-Kader, M. G. and E. Y.-C. Lin (2009). *Performance Measurement of New Product Development Teams*. London, Palgrave Macmillan.

Adams, R., J. Bessant, *et al.* (2006). Innovation management measurement: A review. *International Journal of Management Reviews* **8**(1): 21–47.

Bessant, J. and J. Tidd (2007). *Innovation and Entrepreneurship*. Chichester: John Wiley & Sons.

Bourne, M., J. Mills, *et al.* (2000). Designing, implementing and updating performance measurement systems. *International Journal of Operations & Production Management* **20**(7): 754–754.

Bremser, W. G. and N. P. Barsky (2004). Utilizing the balanced scorecard for R&D performance measurement. *R & D Management* **34**(3): 229–238.

Cedergren, S., A. Wall, *et al.* (2010). A performance evaluation framework for innovation. *Innovation in Business and Enterprise: Technologies and Frameworks*. L. Al-Hakim and C. Jin (Eds). Hershey, PA: IGI Global.

Christensen, C. M. (2003). *The Innovator's Dilemma: The Revolutionary Book That Will Change the Way You Do Business*. New York: Collins.

Cooper, R. G. (1990). Stage-Gate Systems: A new tool for managing new products. *Business Horizons* **33**(3): 44–55.

Cooper, R. G. (2001). *Winning at New Products: Accelerating the Process from Idea to Launch*. Cambridge, MA: Perseus Books.

Cooper, R. G. (2008). Perspective: The Stage-Gate® idea-to-launch process — Update, what's new, and nexgen systems. *Journal of Product Innovation Management* **25**(3): 213.

Cooper, R. G. and E. J. Kleinschmidt (2007). Winning business in product development: The critical success factors. *Research Technology Management* **50**(3): 52–66.

Davila, T., M. J. Epstein, *et al.* (2013). *Making Innovation Work: How to Manage It, Measure It, and Profit from It, Updated Edition*. Upper Saddle River, NJ: Pearson Education.

Driva, H., K. S. Pawar, *et al.* (2001). Performance evaluation of new product development from a company perspective. *Integrated Manufacturing Systems* **12**(5): 368–378.

Eisenhardt, K. M. (1989). Building theories from case study research. *Academy of Management Review* **14**(4): 532–550.

Ernst, H. (2002). Success factors of new product development: A review of the empirical literature. *International Journal of Management Reviews* **4**(1): 1–40.

Ettlie, J. E. and J. M. Elsenbach (2007). Modified Stage-Gate® regimes in new product development. *Journal of Product Innovation Management* **24**(1): 20–20.

Evanschitzky, H., M. Eisend, *et al.* (2012). Success factors of product innovation: An updated meta-analysis. *Journal of Product Innovation Management* **29**: 21–37.

Foos, T., G. Schum, *et al.* (2006). Tacit knowledge transfer and the knowledge disconnect. *Journal of Knowledge Management* **10**(1): 6–18.

Gharajedaghi, J. (2006). *Systems Thinking: Managing Chaos and Complexity — A Platform for Designing Business Architecture*. San Diego, CA: Elsevier.

Goffin, K. and R. Mitchell (2005). *Innovation Management: Strategy and Implementation Using the Pentathlon Framework*. London: Palgrave MacMillan.

Griffin, A. (1997). PDMA research on new product development practices: Updating trends and benchmarking best practices. *Journal of Product Innovation Management* **14**(6): 429–458.

Hertenstein, J. H. and M. B. Platt (2000). Performance measures and management control in new product development. *Accounting Horizons* **14**(3): 303–323.

Jiménez-Zarco, A. I., M. P. Martínez-Ruiz, *et al.* (2006). Performance measurement system (PMS) integration into new product innovation: A literature review and conceptual framework. *Academy of Marketing Science Review* **10**(7).

Kahn, K. B. (2005). *The PDMA Handbook of New Product Development*. Hoboken, NJ: John Wiley & Sons.

Kaplan, R. S. and D. P. Norton (1992). The balanced scorecard measures that drive performance. *Harvard Business Review* **70**(1): 71–80.

Katz, D. and R. L. Kahn (1978). *The Social Psychology of Organizations*. New York: John-Wiley & Sons.

Kerssens-van Drongelen, I. C., B. Nixon, *et al.* (2000). Performance measurement in industrial R&D. *International Journal of Management Reviews* **2**(2): 111–143.

Koehler, C. and R. Weissbarth (2004). The Art of Underengineering. *Strategy+Business Magazine*.

Neely, A. (2005). The evolution of performance measurement research: Developments in the last decade and a research agenda for the next. *International Journal of Operations & Production Management* **25**(12): 1264–1277.

Neely, A. (2007). *Business Performance Measurement*. Cambridge: Cambridge University Press.

Neely, A., C. Adams, *et al*. (2001). The performance prism in practice. *Measuring Business Excellence* **5**(2): 6–12.

Nixon, B. (1997). Conference report: Performance measurements for R&D. *R & D Management* **27**(1): 87–90.

Paranjape, B., M. Rossiter, *et al*. (2006). Performance measurement systems: successes, failures and future — A review. *Measuring Business Excellence* **10**(3): 12.

Patton, M. Q. (2002). *Qualitative Research and Evaluation Methods*. Thousand Oaks, CA: Sage Publications.

Poolton, J. and I. Barclay (1998). New product development from past research to future applications — A study of success and failure in the U.S. electronics industry. *Industrial Marketing Management* **27**(3): 197–212.

Pun, K. F. and A. S. White (2005). A performance measurement paradigm for integrating strategy formulation: A review of systems and frameworks. *International Journal of Management Reviews* **7**(1): 22.

Robson, C. (2002). *Real World Research*. Oxford, UK: Blackwell Publishing.

Rothwell, R., C. Freeman, *et al*. (1974). SAPPHO updated — Project SAPPHO phase II. *Research Policy* **3**(3): 258–291.

Schilling, M. A. (2006). *Strategic Management of Technical Innovation*. New York, NY: McGraw-Hill.

Seider, R. (2006). Optimizing project portfolios. *Research Technology Management* **49**(5): 43–48.

Senge, P. M. (1990). *The Fifth Discipline, the Art & Practice of Learning Organizations*. New York, NY: Currency Doubleday.

Syamil, A., J. Doll William, *et al*. (2004). Process performance in product development: measure and impacts. *European Journal of Innovation Management* **7**(3): 205–217.

Tang, V., B. Liu, *et al*. (2005). Enabling factors in successful product development. *International Conference on Engineering Design*. Melbourne, Australia.

Tatikonda, M. V. (2008). Product development performance measurement. *Handbook of New Product Development Management*. C. H. Loch and S. Kavadias. Oxford: Butterworth-Heinemann.

Ulrich, K. T. and S. D. Eppinger (2008). *Product Design and Development*. Singapore: McGraw-Hill Education.

Wheelright, S. C. and K. B. Clark (1992). *Revolutionizing Product Development: Quantum Leaps in Speed, Efficiency, and Quality*. New York, NY: Free Press.

Yin, R. K. (2013). *Case Study Research: Design and Methods*. Thousands Oaks, CA: Sage Publications.

CHAPTER 12

New Product Development: Design Risk Assessment

Chung Yeh, Wen-Chin Liu† & Chih-Lieh Lee‡*

To cope with uncertainty, risk management had been initiated at the beginning of the new product development (NPD) project. Risk management is the systematic process of identifying, analyzing, and responding to project risk. Failure Modes and Effects Analysis (FMEA) is widely spread and used to manage risks in diversity industrials. The benefit of FMEA concept and methods has a measurable way to prioritize and minimize risk. The chapter provides a similar method to emphasize to focus on the risk assessment and action selection from different aspects of NPD process.

A corporation's NPD project defines the set of customer needs that it seeks to satisfy through its products and services. From the customer's point of view, they concentrate on technological performance, cost, and schedule items of products or services. Therefore, optimization of the design process from elements to a whole system is the best way to meet the

*Associate Professor and Director of the Department of Industrial Engineering and Systems Management, Feng Chia University, Taichung, 407, Taiwan, R.O.C.

†Manager, Aero Industry Development Corporation, 13FL-3, No. 388, Tung Hwa Rd. Pei-Tung Division, Taichung 406, Taiwan, R.O.C.

‡Senior Engineer, Aeronautical Systems Research Division Chung-Shan Institute of Science and Technology, Taichung, 407, Taiwan, R.O.C.

customer's need throughout the NPD project life cycle. Consequently, we aim to solve the problem by using Genetic Algorithm (GA) for the NPD process. The GA technique analyzes and finds a trade-off solution for the NPD project while finding the optimum parameters combination of elements.

Background

Facing intensive global competition, rapid technological change, and shifting global markets, NPD is essential to corporations in planning for their future. To get high revenue, profit, and competitive advantage, firms have to conduct NPD projects more effectively and efficiently. In fact, NPD project's failure rate is relatively high. Many papers discussed that successful NPD requires effective strategies for reducing risk. Knowledge management systems have the potential to aid in risk reduction (Cooper, 2003). Product development is seen as "among the essential processes for success, survival, and renewal of organizations, particularly for firms in either fast-paced or competitive markets" (Brown and Eisenhardt, 1995). Project teams performing NPD face pressure to reduce cycle time and development costs, without sacrificing innovation, as characterized by a faster–better–cheaper philosophy (McDonough, 2000). Likewise, there are lots of high-level uncertainties inherent during the product development process that project managers and team members must manage uncertainty.

A corporation's NPD project defines the set of customer needs and seeks to satisfy them through its products or services. It is useful in identifying the resources needed and choosing appropriate response actions to control, and manage the identified risk factors in achieving the desired results of the project (Tummala and Mak, 2001). Since there are numerous choices to design and assemble a system from elements and subsystems design at concept or preliminary design stage, it is best then to optimize the design process from elements to whole system to best fit customer's needs throughout the NPD project life cycle.

To cope with uncertainty, risk management had been initiated at the beginning of the NPD project. Design process consists of a series of divergent and convergent activities at varying levels from concept formulation to detailed design (Couger, 1996). Risk management can be approached as a multiphase "risk analysis" which covers identification, evaluation, control, and management of risks (Cooper and Chapman, 1987). Risk management is the systematic process of identifying, analyzing, and responding to project risk. FMEA is widely used in the design, process, and machinery phases of a project by firms. SAE J1739 (Warendale, 2002) practice described FMEA is an evaluation process for analyzing and assessing failures in a system. The objective is to determine the effect of failures on system operation, identify the failures critical to operational success and personnel safety, and assess each potential failure according to the effects on other portions of the system. The design FMEA is used to aid in the objective evaluation of the design, and to evaluate the initial design for manufacturing and assembly service. It supports the design process in reducing the risk of failures. Risk Priority Number (RPN) is one of the FMEA techniques used to identify the individual systems, subsystems, and/or elements of FMEA ranking order. RPN is the product of the severity, occurrence, and detection ranking.

The model then becomes more complex due to the scores of severity, occurrence, and correlation group from individual technological performance, cost, and schedule items. There are several types of algorithms about heuristic improving search algorithm. Three of the most popular are: tabu search, simulated annealing, and GA (GA methodology aimed at integrating life cycle costs studies with risk assessments to provide computational tools for decision support systems) (Gayraud and Singh, 1997). The GA is the fastest algorithm to find optimum solution for the traveling salesman problem (Rardin, 1998), GA has also been successfully used for engine-fault diagnostics. (Ogaji *et al.,* 2002). Rule-based GA associated with embedded project scheduling to solve the risk-based partner selection in a virtual enterprise problem (Ip and Huang, 2003). Therefore, the optimization technique based on the GA was found to be the most suitable for many kinds of problems. The GA technique analyzes and finds the best solution for the NPD project while finding the optimum parameters combination of elements. Consequently, we try to solve the problem by using GA for the NPD process.

Research Methodology

Firms have no choices in conducting NPD through innovation to enhance product added values. From issues of project objective, conceptual design, preliminary design, detail design, production to system utilization, every process has to follow customer's requirements. Rapid and punctual NPD has become a top priority in many organizations as competitors rush to commercialize emerging technologies and to satisfy customer needs (Roberto *et al.*, 2004).

In order to follow customer's requirements, a project has to issue same or higher target of risk control. With the high level of uncertainty inherent during NPD design process, project team members often give more effort to risk management. NPD risk evaluation consists of the process of risk identification, qualitative risk analysis, and quantitative risk analysis.

We have developed a set of risk criteria for the NPD design stage. It contains technological performance, schedule, and cost. The work breakdown structure (WBS) provides the control structure by which all project works are managed, evaluated, and reported. Every work can be identified and defined by the WBS that specified cost, schedule, and performance requirement. For example, the scheduling process is fully integrated throughout the WBS by a logic network that establishes interdependency between WBS elements. The element, subsystem, and system schedules then become the framework for establishing time-phrased estimates and resource requirements to meet contractual obligations.

To establish working scope, budget, and schedule baselines that is integrated at the individual level and upward through all levels of the NPD WBS. Measuring, processing, and reporting the risk management status that against technology work scope, budget, schedule baselines to the appropriate levels of internal and external management.

Potential failure mode is used to define the function of the system, subsystem, or element fail in the life cycle for FMEA process. It is not only suitable for technical risk, but also suitable for cost and schedule risk between the concept and preliminary design stage. Therefore, Project Management Institute stated technique, cost, and schedule are triple constraints for an NPD project, and stakeholders have to handle and find an optimum characteristic of these elements (Project management Institute, 2004).

The severity and occurrence definitions follow the FMEA to define the elements' specifications. Severity is the rank associated with predicting the worst possible result versed planned result for a given item. Severity is the relative ranking score depending on the evaluation criteria of the individual item. The criteria ranking values are from 1 to 10 for severity and occurrence items. Occurrence is the likelihood that the predicted possible result will occur during the product life. The likelihood of occurrence ranking number has a relative meaning rather than an absolute value. Detection is the rank associated with best detection control for a given item in the FMEA process and will be utilized at the system definition stage. We adopt correlation to replace the detection at preliminary design stage.

A correlation system is a method to quantify relative deliverable status of technique, cost, and schedule for work. The technique and delivery of a work is determined by the resource being used during the NPD stage (Feng *et al.*, 2000).

It is reasonable to assume that triple constraints of a working item correlated. It is more complicated to describe the importance or relation between the technique, cost, and schedule during the design process. Just like a weighting system, it will assign a numerical weight to each of the evaluation criteria, and get the whole score for the three items. The criteria are ranked from 1 to 10 for the correlation model. Sum of the three items correlation number is then equal to one. The criteria and scores are defined by the project team which references the FMEA definition. These numbers came out from previous project and common consensus of team members. So, project can define the criteria to fit in the NPD characteristics.

The model used on the three core items of technological performance, cost, and schedule risk to improve the process of preliminary design and resource allocation in NPD projects. When two parts are assembled into one assembly, it encodes the individual score from different view as decision variables in a string of real number digits. One part has 1×10^9, the product of 1000 combination by severity, occurrence, and correlation score — 10 combination for each element' technological performance, cost, and schedule, combinations for the technique, cost, and schedule item that have individual severity, occurrence, and correlation system choices. The other assembled part also has 1×10^9 combinations. Therefore one assembly, which has two interdependent parts assembled, totaling 1×10^{18} combination, is too complicated to choose best design configuration. Therefore, we adopted the GA to search the best combination of respective parameters. GA has powerful optimization tools of selection, and mutation, and crossover techniques by combining the improved solution of initial solution. GA is capable to seek an optimal combination of short manufacturing time and others resources (Jou, 2005). The initial concept design of two related elements is a feasible initial solution for the NPD project, and the two related elements are two interdependent populations at design stage.

The operations of selection, and mutation, and crossover techniques are as follows:

- Selection: Each new generation choose the best solutions held over from previous generations for two individual populations.
- Mutation: Each new generation of immigrant solutions added randomly increase diversity from previous generations for two individual populations.
- Crossover: A pair of new generation chromosomes produced from randomly combining and reassembling from a pair of selected previous generations at same point.

GA calculates over a number of iterations and each of these iterations is based on three fundamental operations. The new iterations are created by three operations to reassemble pairs of individuals in each population. Designer and manager evaluate the new interdependent generation's characteristic to review the goal of NPD project.

Customers expect and desire to get low price, high quality, and short time to market an NPD product. Therefore, NPD project management is designed to manage or control firm resources on a given activity, within time, cost, and performance of firm or customer. Kerzner identified time, cost, and performances as the constraints on the project (Kerzner, 2003). The proposed flowchart of an NPD cycle is shown in Fig. 1. It would create lots of conflict while team members release the detail design. Project management should focus to confront the conflict in front of detail design stage. We can allocate the optimum resources during the preliminary design stage. We will get the minimum design change at the detail design and production stage.

The FMEA technique is adopted during detail design, production, and system utilization stage by the firm. At the preliminary design stage, the project team members have initiated the analysis model. The cost, schedule, and technology risk are identified from history or expert information/data. Tradeoff of resources depends on the parameters of elements' characteristics.

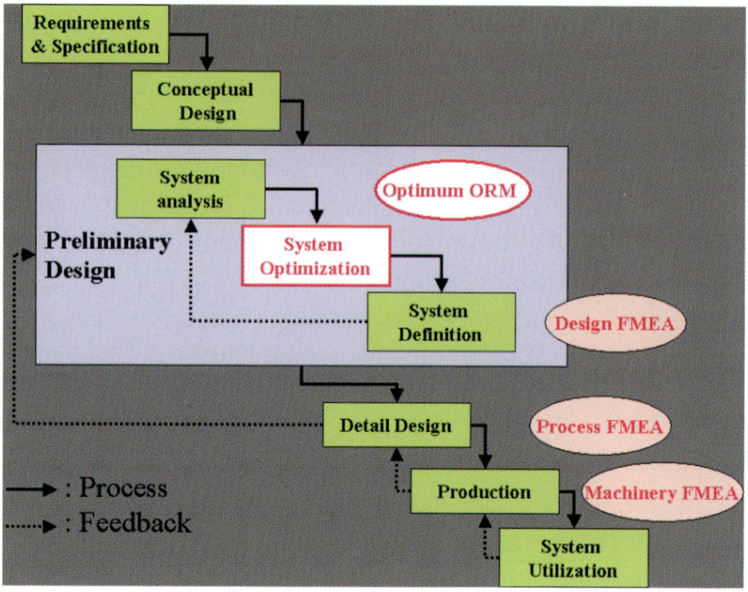

Figure 1: Proposed flowchart of NPD cycle.

Model Building and Software Implementation

Development of Risk Evaluation Model

We first establish the risk handling process for the risk analysis model at risk management planning stage. Follow the previous approaches to the problems in the Research methodology section. Then, the risk categories of technology, schedule, and cost are built at risk identification stage. Third, the risk score of severity, occurrence, and correlation are evaluated for risk categories. Fourth, corrective and preventive actions are made at the risk response stage. The evaluation of severity has impact, criteria, and score for technology, schedule and cost items, as shown in Table 1.

The evaluation of occurrence has probability, criteria, and score for technology, schedule, and cost items, as shown in Table 2.

Table 1: Evaluation criteria of severity.

Impact	Criteria			Score
	Technology	Schedule	Cost	
Very high	Very high severity when a failure mode affects major performance	>6 months extended of estimated work duration	>8% cost added of estimated work budget	10
		Up to 6 months	Up to 8%	9
High	System inoperable or reduce the level of performance	<6 months and ≥5 months	<8% and ≥7%	8
		<5 months and ≥4 months	<7% and ≥6%	7
Moderate	System operable or customer unsatisfied	<4 months and ≥3 months	<6% and ≥5%	6
		<3 months and ≥2 months	<5 and ≥4%	5
		<2 months and ≥1 month	<4% and ≥3%	4
Minor	Minor nonconformity but not reduce the level of performance	<4 weeks and ≥3 weeks	<3% and ≥2%	3
		<3 weeks and ≥1 week	<2% and ≥1%	2
None	No distinguishable effect	No impact	No impact	1

Note: The criteria are defined by the project team. These numbers came out from previous project and common consensus of team members.

Table 2: Evaluation criteria of occurrence.

	Criteria			
Probability	Technology	Schedule	Cost	Score
Very high	Persistent outcome, >50%			10
	Up to 50%			9
High	Frequent outcome, ≥40% and <50%			8
	≥30% and <40%			7
Moderate	Occasional outcome, ≥20% and <30%			6
	≥10% and <20%			5
	≥5% and <10%			4
Low	Relatively few outcome, ≥3% and <5%			3
	≥0.01% and <3%			2
Remote	Outcome is unlikely, <0.01%			1

Note: The criteria are defined by the project team. These numbers came out from previous project and common consensus of team members.

The evaluation of correlation has correlation, criteria, and score for technology, schedule, and cost items, as shown in Table 3.

In Tables 1–3, we define the risk criteria ranking score for NPD project risk analysis. The severity, occurrence, and correlation risk scores are evaluated for technology, schedule, and cost categories. The column is used to classify any category for system, subsystem, or element that requires evaluation. The score of individual impact status depends on predicting the worst possible result. We find that the score and classification risk criteria ranking values are consistent for every element in the NPD project.

The risk criteria are standardized for the NPD project team members who are familiar with FMEA evaluation technique. According to the NPD process in Fig. 1, team uses the risk management model in preliminary design and FMEA in detail design, production, and so forth. So, the evaluation process is integrated and consistent for NPD life cycle. At preliminary

Table 3: Evaluation criteria of correlation.

Correlation	Criteria: Deliverable Technology	Meet Schedule	Acceptable Cost	Score
Very high	Critical technology under development or transfer	Affect to critical schedule >3 months	>10% additional target budget	10
		Up to 3 months	Up to 10%	9
High	Single source but adoptable or received	<3 months and ≥2 month	<10% and ≥9%	8
		<2 months and ≥1 month	<9% and ≥8%	7
Moderate	Multi source or upgraded technology	>3 months slip in key milestone	<8% and ≥7%	6
		≤3 months and >2 months	<7% and ≥4%	5
		≤2 and >1 month	4% and ≥3%	4
Low	Revised or minor upgraded technology	More than 5% additional allocation resources to meet schedule	<3% and ≥2%	3
		>2% and ≤5%	<2% and ≥1%	2
Remote	Original technology	No impact	Less than 1%	1

Note: The criteria are defined by the project team. These numbers came out from previous project and common consensus of team members.

design stage, each score of time, cost, and performance are evaluated for elements, subsystems, and systems.

Model Building

The risk analysis model can be described as follows:

Each element has different severity, occurrence, and correlation score by the technology, cost, and schedule category. For element i, $i = 1, 2, 3 \ldots i$. For risk category j, $j = 1$ (technology), 2 (cost), and 3 (schedule). For generation number n, $n = 1, 2, 3 \ldots n$. The i, j, and n, which are positive integer in the model.

The corresponding decision variables are:

$S_{i,j} \equiv$ score of severity item j to be produced at element i

$F_{i,j} \equiv$ score of occurrence item j to be produced at element i

$w_{i,j} \equiv$ score of correlation item j to be produced at element i

$W_{i,j} \equiv$ score of correlation matrix $1{\sim}j$ to be produced at element i

$x_{i,j} \equiv$ score of risk item j to be produced at element i

$X_i \equiv$ score of overall risk item matrix $1{\sim}j$ to be produced at element i

$G_i(n) \equiv$ overall risk score of generation n of element i

We will to minimize some combination of risk score difference and risk ranking to select the best element 1 and 2. Then a multivariable nonlinear programming model is:

Minimize $|\,(G_1(n) - G_2(n))\,|$ (risk score difference) (1)

Minimize $G_i(n)\,|$ risk ranking) (2)

Setup constraints and calculation

$$x_{i,j} = S_{i,j} \times F_{i,j}$$
$$X_{i,j} = [x_{i,1}\,, x_{i,2}\,, x_{i,3}\,] = [S_{i,1} \times F_{i,1}\,, S_{i,2} \times F_{i,2}\,, S_{i,3} \times F_{i,3}]$$
$$W_{i,j} = [w_{i,1}, w_{i,2}, w_{i,3}]^T$$

$$\sum_{(j=1\sim3)} w_{i,j} = 1\ , i = 1\sim i$$

$$W_{i,j}^{(n+1)} = \frac{W_{i,j}^{(n)}}{W_{i,1}^{(n)} + W_{i,2}^{(n)} + W_{i,3}^{(n)}}, \mathrm{i} = 1\sim n,\ j = 1\sim3$$

$$G_i = \sum_{(j=1\sim3)} X_{i,j} \times W_{i,j}$$

The limitation of associated decision variables:

$$1 \leqq S_{i,j} \leqq 10$$
$$1 \leqq F_{i,j} \leqq 10$$
$$1 \leqq w_{i,j} \leqq 10$$
$$0 \leqq |W_{i,j}| \leqq 1$$
$$1 \leqq x_{i,j} \leqq 100$$
$$1 \leqq G_i(n) \leqq 100$$

Where $S_{i,j}$, $F_{i,j}$, $w_{i,j}$ are positive integer between 1 and 10.

GA method for combining solutions of the population is known as selections, mutations, and crossovers. Each generation of individual elements has parent characteristics or genes. Initial design for every element is an initial solution for the system. The system's best solution will appear in the part of the population.

In general, all solutions will be feasible, and some might be as good as global optimum in the population. Combining pairs of parent generations creates new solutions. The method is initiated from a pair of parent solutions, and then sequentially produced a pair of children solutions by elites, immigrants, and crossovers actions.

Two initial solution of two elements design characteristic are:

$$X_1 = [x_{1,1}, x_{1,2}, x_{1,3}] = [S_{1,1} \times F_{1,1}, S_{1,2} \times F_{1,2}, \mathrm{S}_{1,3} \times F_{1,3}]$$
$$W_1 = [w_{1,1}, \mathrm{w}_{1,2}, \mathrm{w}_{1,3}]^T$$
$$X_2 = [x_{2,1}, x_{2,2}, x_{2,3}] = [S_{2,1} \times F_{2,1}, S_{2,2} \times F_{2,2}, \mathrm{S}_{2,3} \times \mathrm{F}_{2,3}]$$
$$W_2 = [w_{2,1}, w_{2,2}, w_{2,3}]^T$$

We can illustrate the overall risk score of generation 1 of element 1, 2:

$$G_1(1) = X_1 \times W_1$$
$$G_2(2) = X_2 \times W_2$$

The value of overall risk score is 1–100 that can be expressed as the element risk ranking in the whole system. The 100 generations are defined to judge the trend of convergence of objective function. We are focused on minimizing the risk score of element 1 and 2 and the difference of the two elements. The higher overall risk score calculated in the system, the more proactive corrective action is needed for the design of the element.

GA strategy is employed to select pairs of current solutions to produce new ones via crossover and immigrant. Therefore, we first decide which new or old solutions will survive in the next generation, and how to maintain diversity in the population. We choose minimum overall risk score of old generation element as new generation genes.

- Selections: Choose the individual minimum $G_i^{(n)}$ item in the 2 element populations. Selected element 1, 2 has minimum overall risk score characteristic to new generation as a mixture of elite solution in the element 1, 2 populations.
- Mutations: Randomize to decide which category added arbitrarily to increase diversity. We use random variable to choose selected element 1, 2 categories that produces one new immigrant feasible solution in the next generation.
- Crossovers: Randomize to decide which selected element 1, 2 items break and reassemble chromosomes of selected element 1, 2 that produces two new crossover feasible solutions in the next generation. The break and reassemble chromosome at the same item of selected element 1, 2.

The next generation can be illustrated with two feasible solutions as:

$$
\begin{aligned}
G_1^{(2)} &= X_1^{(2)} \times W_1^{(2)} = S_{1.1}^{(1)} \times F_{1.1}^{(1)} \times W_{1.1}^{(1)} \\
W_1 &= [w_{1,1}, w_{1,2}, w_{1,3}]^T \\
X_1^{(2)} &= [x_{2,1}, x_{2,2}, x_{2,3},] = [S_{2,1} \times F_{2,1}, S_{2,2}, \times F_{2,2}, S_{2,3}, \times F_{2,3},] \\
W_2 &= [w_{2,1}, w_{2,2}, w_{2,3}]^T
\end{aligned}
$$

The calculation flowchart of risk analysis model from element characters input, and parameter of the model setup, objective function calculate, generation less than 100 check, selection implement, crossover implement, and mutation implement, as illustrated in Fig. 2. In the model, we expect to evaluate the risk trend of new generations, so 100 generations are defined to judge the trend of convergence of objective function. We then focused on minimizing the risk score of element 1 and 2 and difference of two elements. The higher overall risk score calculated in the system, the more proactive corrective action is needed for the design of the element. Parameters of risk analysis model have to commit the characteristics of the project. There are several parameters that can be corrected: generation, objective function, population, selection condition, crossover condition, mutation condition, and constraints.

Software Implementation

The Overall Risk Management model was developed on a Borland Delphi 6, based on Microsoft Window platform. Delphi was and still is one of the most popular and best object-oriented programming and visual programming for Windows. Delphi is adapted to supports database applications in the programming environment (Cantu, 2001). Therefore, we can easily write data-access code for the populations.

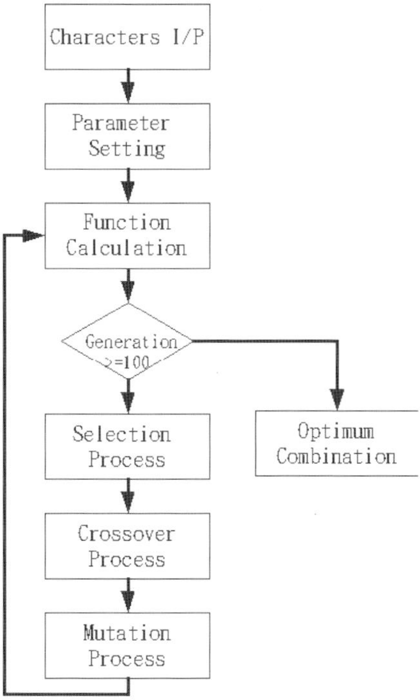

Figure 2: Calculation flowchart of integrated risk analysis model.

The model consists of two work forms within a single Delphi execution file and a Borland Database Engine originated with Paradox file. The model attempts to capture the respective element's characteristic using cost, schedule, and performance requirement. We first build up the interdependent WBS elements relation for identification. There are two work forms in the software. Work form 1 is intended to input the two element's parameters that establish interdependency between WBS elements. Each WBS element is evaluated the same severity, occurrence, and correlation risk scores evaluation criteria of its respective items. By scrolling down the individual dialog box, the user can easily input nine items value for individual element and get eight outputs as in Fig. 3.

User can check the output values directly from the work form 1 to make sure the design concept is conformed to the

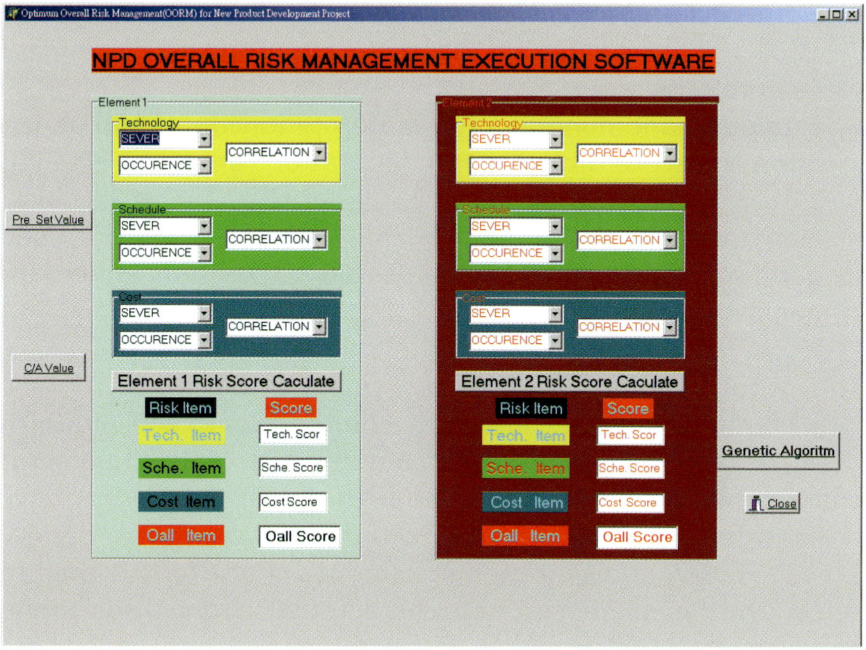

Figure 3: Risk scores of element 1 and 2 inputs of integrated risk analysis model.

design requirement. The risk scores are expressed as cost, schedule, performance, and overall risk status for WBS element. The output values can help the designer of two elements to think over and redesign the interface and function between two elements. If the result of work form1 is acceptable, the user can go to work form 2 for GA calculations.

In work form 2, user input two WBS-element parameters input database as parent chromosomes for GA technique search. The correlation characteristic has been merged with occurrence characteristic for simplification. It will generate the element 1, 2 risk score output values:

- Element 1, 2, and minimum risk score of individual child generation risk score table for 100 generation ranked by generation

- Element 1 and minimum risk score of individual child generation risk score table for 100 generation ranked by element 1 overall risk score
- Element 2 and minimum risk score of individual child generation risk score table for 100 generation ranked by element 2 overall risk score
- Element 1, 2, and minimum risk score of individual child generation risk score chart for 100 generation ranked by generation

The output table and chart are to help user to clarify the two elements possible and best combinations. The model is designed to choose optimum combination of two elements characteristic without any constraints. In fact, designers have to consider lots of restrictions from the real world. So, designers can review and change the element's characteristic then to select another element's best characteristic from three tables and a chart.

Case Application — The Turbo Fan Engine Development Project

Engine Electronic Control

Overall Risk Management model was developed from model building and software implementation. To address the model effectively, we arranged for a field test at a turbo fan engine development department in Taichung, Taiwan. The department tried to assess the risk of engine electronic control (EEC) development project using the risk evaluation model.

EEC is the primary interface between the engine and the aircraft. The pilot controls the throttle and fuel switches in the cockpit, then EEC receives the signals from throttle, and fuel switches position, and engine information, and flight conditions and delivers several outputs to control the engine. EEC improves the efficiency of engine operation for different flight condition in the following manner (Treager, 1997):

- Enhancing engine control functions
- Limiting engine operational area
- Improving engine operational reliability
- Improving engine maintenance
- Improving communication between engine and throttle

Followed process SAE APR1834 (Warrendale, 1997), EEC which consists of central process unit board and input/output board runs the fault/failure analysis to cross-check the design requirements during preliminary design stage. EEC design team, around 12 persons that includes project leader, software designers, hardware designers, configuration engineer, QC person, customer representative, buyer, potential venders, ran the model to tune the elements' perspectives about risk assessment prior to fault/failure analysis.

Data Input and Result

The EEC design team worked together and input the original design configuration from different aspects of cost, time, and technological performance as a preference resource allocation. Team defined the criteria and score by previous project and common consensus of team members. Element 1 is EEC central process unit board and element 2 is EEC input/output board. By scrolling down the nine dialog box and push down "risk score calculate" bottom, as illustrated in Fig. 4.

In Fig. 4, the risk score of element 1(EEC central process unit board) is higher than element 2 (EEC input/output board). The overall risk score is 50.5, technology and schedule risk score are 56 for element 1, design team didn't think it is a good combination for EEC assembly. It means the new EEC assembly system risk is too high to execute the detail design of central processing

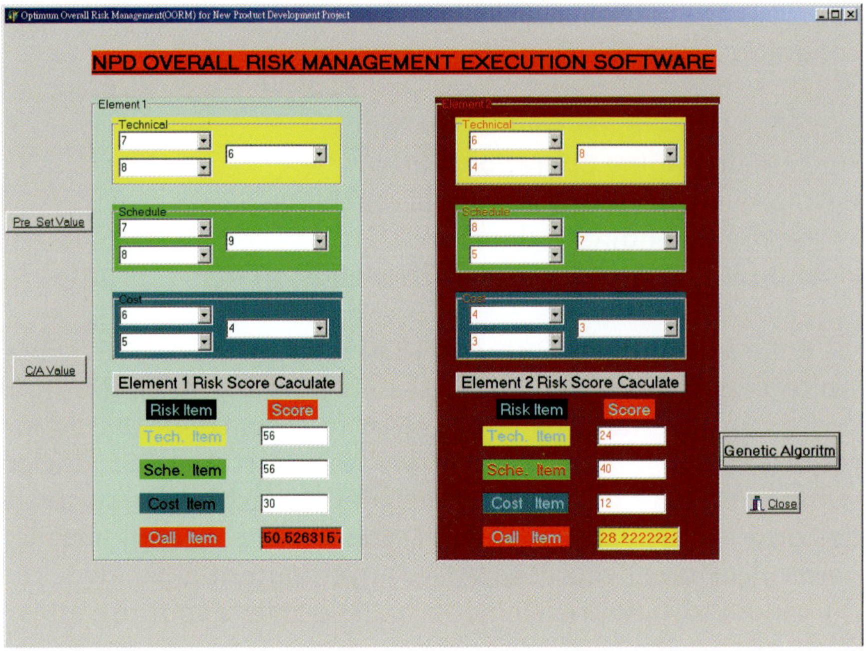

Figure 4: Input parameters of element 1 and 2 of integrated risk analysis model.

unit (CPU) board and input/output (I/O) board. After dealing with team members' capability and assessing uncertainties, they concluded another design configuration. Team tried to find possible combination as continuous improvements for next 100 generations about EEC assembly design improvements using GA. The result illustrated in Fig. 5. We can see the trend of the reproducing process from the chart. The minimum risk score is 26.98 for element 1 at generation 48 in element 1 score ranking table. The minimum risk score is 26.98 for element 1 at generation 90 in element 1 score ranking table. The software model is allowed to move the scroll of each table to check every items value.

The EEC team reviewed the result and found better solution in the preliminary design review. Then, they issued the proposal

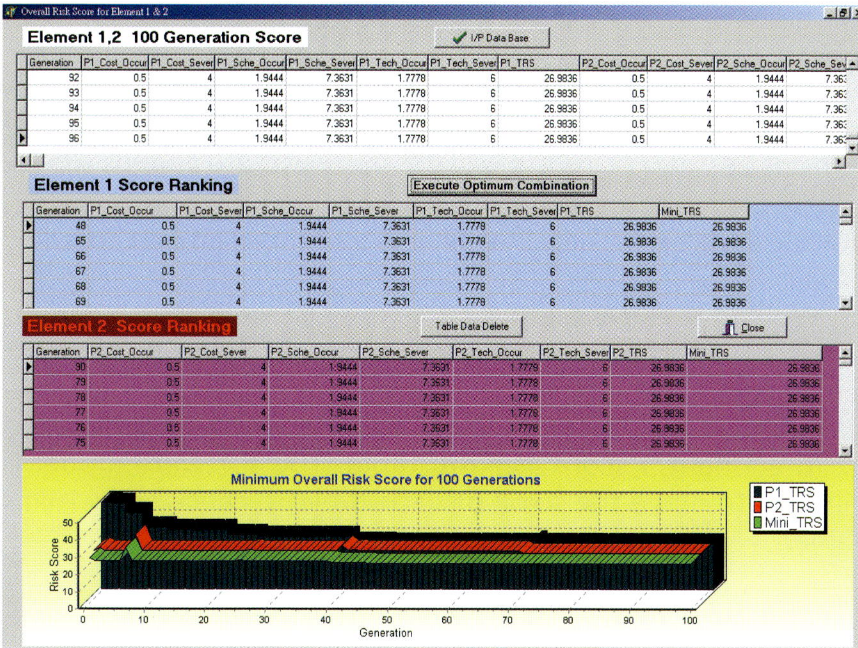

Figure 5: Results of element 1 and 2 of integrated risk analysis model.

to project manager about resources arrangement to reduce downstream risk. EEC team suggested adopting new technology (original severity of 7 to be of 7, original occurrence 8 to be of 6), early schedule (original severity of 7 to be of 6, original occurrence 8 to be of 5) but higher cost (original severity of 6 to be of 7, same occurrence of 5) for CPU board and higher specification (original severity of 6 to be of 7, original occurrence 4 to be of 3) and early schedule (original severity of 8 to be of 6, same occurrence of 5) in I/O board. The cost to complete is higher than previous plan. So, the correlation values of two elements are fixed, the severity and occurrence values of two elements changed by corresponded condition.

The risk score of schedule item for both elements are decreased acutely from 56 to 30 for element 1, from 40 to 30 for element 2. The risk score of technology item for both elements also decreases, too. The only one risk score increase which is element 1 cost item increasing from 30 to 35. The overall risk score decreases from 50.5 to 34.8 for element 1, from 28.2 to 23 for element 2. In comparison with others systems, the EEC is lower risk than others. It represents the overall risk score of new EEC assembly reduced and accepted by project.

We can see the trend of the downstream improvement process from the chart. The minimum risk score is 20.37 for element 1 at generation 76 in element 1 score ranking table. The minimum risk score is 20.42 for element 1 at generation 97 in element 1 score ranking table. The software model is allowed to move the scroll of each table to check every items value.

Discussions and Summary

NPD project risks include cost, budget, and schedule, and quality, and/or technological performance, and contract relationships, and so forth. Project members can assess the risk of the element during NPD life cycle. We can see the different design concept deliver various level of risk for EEC assembly. The original and revised risk identification was shown in Table 4. Project team can have the risk sense to discuss and revise the function of elements by the interdependent elements overall risk score.

It is very important that risk assessment is established as early as possible in the NPD project. The aim of team is to reduce or eliminate the design change of implementing preventive and corrective actions. The project members can tradeoff between the various objectives and reallocate between the resources for optimum concept design. By the way, designers can focus on the worst condition from the risk assessment to refine the design and monitor others conditions. If optimum

Table 4: Risk scoring summary sheet of integrated risk analysis model.

WBS Number	WBS Item	Configuration	Risk Score Technical	Schedule	Cost	Overall
0001	CPU Board	Original	56	56	30	50.5
		Revised	42	30	35	34.8
0002	I/O Board	Original	24	40	12	28.2
		Revised	21	30	12	23
0010	EEC Assy.	Original	10.7/ 17.6	14.3/ 26.5	2/ 6.3	27/ 50.5
		Revised	8.6/ 13.2	9.7/ 14.2	2/ 7.4	20.4/ 34.8

design performance is to be achieved, the assembly element should be a minimized risk score or less modification.

In this case, we adjusted the parameters of risk analysis model to be used within the characteristics of the project. There are several parameters that can be corrected: generation, objective function, population, selection condition, crossover condition, mutation condition, and constraints.

Our experience and the engine development department suggested that it is not possible to predefine a universal set of risk parameters without historical information and expert experience. The criteria and parameters of model should be redefined for different NPD project and firms characteristics. The risk evaluation model is not valid for all projects with the same criteria and parameters. Moreover, the current model lacks of the flexibility to setup corresponding parameters with more than two interdependent WBS elements to calculate the overall risk score.

Conclusions

The risk analysis model is a data-based tool with the causes and quantified effects from a global technology, schedule, and cost perspectives. It attempts to help NPD project team members estimate risk margins for proposals that is submitted to project member, executive, or customer. Moreover, it can build a flexible guideline that permit rapid change in element configuration.

The model contains several useful features that NPD can be used implement risk analysis. It can supply the technological performance, schedule, cost risk score, and overall risk score for specific interdependent elements. Project member allocates the resources at preliminary design stage in lieu of design change at detail design and/or production stage. Optimal combination of specific interdependent elements can be modified at the beginning of initial design stage.

Past experience has shown that FMEA and GA are the most widely used and well accepted by firms. We conclude that the risk analysis model can be viewed as risk simulation within NPD project implement. The model can incorporate FMEA technique to qualitative and quantitative risk analysis of NPD product.

Acknowledgements

The authors gratefully acknowledge the financial support of National Science Council R.O.C., (Grant No. NSC 95-2221-E-035-088) and Aeronautical Systems Research Division Chung-Shan Institute of Science and Technology supported this work.

References

Brown, S. L., and Eisenhardt, K. M. (1995). Product development: Past research, present findings, and future directions. *Academy of Management Review*, **20**: 343–378.

Cantù, M. (2001). *Mastering Delphi 6*. Alameda, CA: Sybex.

Cooper, D., and Chapman, C. (1987). *Risk Analysis for Large Projects-Models, Methods and Cases*. Chichester, UK: John Wiley and Sons.

Cooper, L P. (2003). A research agenda to reduce risk in new product development through knowledge management: A practitioner perspective. *Journal of Engineering and Technology Management*, **20**: 117–140.

Couger, J. D. (1996). *Creativity and Innovation in Information Systems Organizations*. New York: Boyd and Fraser.

Feng, C. W., Liu, L., and Burns, S. A. (2000). Stochastic construction time-cost trade-off analysis. *Journal of Computing in Civil Engineering*, **14**: 117–126.

Gayraud, S., and Singh, R. (1997). A techno-economic computational tool for power generation project assessments and life cycle risk management. *ASME Turbo Asia*, 147–150.

Ip, W. H., and Huang, M. (2003). Genetic algorithm solution for a risk-based partner selection problem in a virtual enterprise. *Computers and Operations Research*, **30**: 213–231.

Jou, C. (2005). A genetic algorithm with sub-indexed partitioning genes and its application to production scheduling of parallel machines. *Computers and Industrial Engineering*, **48**: 39–54.

Kerzner, H. (2003). *Project Management: A Systems Approach to Planning, Scheduling and Controlling*, 8th ed. New Jersey: John Wiley and Sons.

McDonough, E. F. (2000). Investigation of factors contributing to the success of cross-functional teams. *Journal of Product Innovation Management*, **17**: 221–235.

Ogaji, S., Sampath, S., and Singh, R. (2002). Novel approach for improving power-plant availability using advanced engine diagnostics. Applied Energy, **72**: 389–407.

Project Management Institute. (2004). *PMBOK Guide: A Guide to the Project Management Body of Knowledge*, 3rd ed. pp. 8–25.

Rardin, R. L. (1998). *Optimization in Operations Research*. Upper Saddle River, NJ: Prentice Hall.

Roberto, F., Luigi, S., and Paolo, T. (2004). Product development time performance: investing the effect of interactions between drivers. *Journal of Product Innovation Management*, **21**: 199–214.

Treager, I. E. (1997). *Aircraft Gas Turbine Engine Technology*, 3rd ed. Boston: McGraw-Hill, pp. 292–300.

Tummala, V. R., and Mak, C. L. (2001). A risk management model for improving operation and maintenance activities in electricity transmission. *Journal of the Operational Research Society*, **52**: 125–134.

Warrendale. (1997). Fault/failure analysis for digital systems and equipment. Society of Automotive, *SAEAPR1834*, USA.

Warrendale. (2002). Potential failure mode and effect analysis in design (Design FMEA), potential failure mode and effect analysis in manufacturing and assembly processes (Process FMEA) and potential failure mode and effect analysis for machinery (Machinery FMEA), *SAEJ1739*, USA.

Chapter 13

Technology Diffusion: Application Paths

Kensuke Kato & Kumiko Miyazaki†*

Most manufacturers are fully aware of the need to develop new businesses based on competitive technologies, thus they are actively exploring business applications of their accumulated expertise in the firms. They build up efforts for enhancing technology commercialization, for example, new energy technologies to solve environmental problems, or security technologies to contribute to safety and a comfortable society. Meanwhile, in the case of developing new businesses based on emerging technologies which do not exist so far, the anticipated markets for the application do not arise as expected at all. For example, even though the markets of mobile fuel cell, optical interconnection, or carbon nanotube are expected to grow rapidly within the next few years, they have kept still almost at the same position.

What are the deadlock factors preventing new business development? The inhibitory factors for the progress are not simply due to technological development issues. Other issues should also be considered, such as further progress of competitive technologies, infant progress of necessary complementary technologies, and shortage of social requirements. The attainment of the business objectives is difficult for a firm to solve by

*EffecTech Institute, Inc., Tokyo, Japan and Stem Cell & Device Laboratory, Inc., Kyoto, Japan.
†Graduate School of Innovation Management, Tokyo Institute of Technology, Tokyo, Japan.

itself in the short-term. In order to overcome the deadlocks, there are not only technical challenges such as upgrading performance, decreasing cost, and developing new functions. Yet, most engineers try to develop technologies as if they are the most important factors. In addition, the issues sometimes include cultural or customary habitual matters such as changing development style from conventional method especially in the case of introducing process innovations.

Thus, what is the effective approach to make progress of a new business? In this paper, an approach intended to overcome or bypass a deadlock in technology commercialization during the introductory phase is examined. In particular, the case of mobile fuel cell called direct methanol fuel cell (DMFC) for product innovation and the case of fluid microelectromechanical systems (MEMS) called microreactor for process innovation are analyzed.

Based on these analyses, it has been shown that since the major inhibitory factors are difficult to be solved in the short term, the application markets should be modified. Therefore, establishing and modeling "paths to create a final market" is crucial not only for commercialization but also for accumulating knowledge, building technology competences, and anticipating social requirements with increasing reliability.

Approach for New Business Development

Previous Literature

Necessity, difficulty, and its appeal for new business development

In order to exploit new developed technologies, firms should move forward in the process of entering or creating a new business in the new market. In this field, a wide range of research such as entrepreneurship, market analyses, and R&D deployment have been carried out so far (e.g., Drucker, 1985; Utterback, 1994; Veldhuizen *et al.*, 2006).

Among these, developing a new business in a market which does not exist, or creating or exploiting emerging market is recognized as a difficult issue. To create value by providing unprecedented new functions or new processes to society is defined by Schumpeter as "new combinations". He stated that "carrying out new combinations is a special process and the object of a special kind of function" as for the difference between execution of a plan for a customary business (normal flow) and an implementation of a new business (innovation), Schumpeter noted that there are enough differences between walking on a road and making progress while making a road (Schumpeter, 1934).

New business development is considered a difficult process but then again its appeal is also considerable. "In development of an emerging market, competitions of technological and business development are caused by pioneers who believe that it is better to be a first mover for prior benefits than a follower" (Lieberman and Montgomery, 1988).

Collaborating with "lead users" and "open innovation"

In many cases of providing new products or services with emerging technologies, requirements from the users will be taken into account, in order to complement the shortage of requirements which are considered inside the firm. Freeman explained quantitatively that understanding the requirements of users or customers by engineers has a high correlation with commercializing innovation (Freeman and Soete, 1974). Both von Hippel and Day pointed out that the possibility of new business start-up can be improved by discovering, collaborating, and learning from advanced lead users who have already faced problems, and attempted to solve them (Day, 2000; von Hippel, 1988).

These advanced lead users are highly motivated due to the expectation of large returns, and are willing to participate in joint development or market research activities for their early markets. In addition, they understand the value of the initial version and tend to have tolerance to accept it even for a high price (Day, 2000).

Here, valuable feedback about the initial market can be provided by the lead users as an advanced indicator. When developing a business, collaborating with users is not only a method for promoting a new business. In the manufacturing industry, materials or components suppliers have a direct or large influence on cost, quality, technology, and time to market. Therefore, importance of building effective alliance with them is increasing (Handfield *et al.*, 1999).

Especially when creating a new business, there are many issues that are difficult to be solved by a single company by itself. As possible approaches in this case, Chesbrough stressed the importance of the following (Chesbrough, 2003); "To find, understand, and filter external knowledge, to develop internal knowledge that is insufficient by external knowledge, to integrate internal and external knowledge, to create a new system, to utilize research achievements which cannot be taken advantage internally to make profit outside the company".

The above points have been recognized by many managers as a basic approach to business success. Recently, collaborative research among universities, public research institutes, and companies are being promoted in Japan.

Deploying technology and path dependency

Furthermore, Miyazaki demonstrated another method to enhance competitiveness by finding new applications through technological competence building (Miyazaki, 1994). In the early phase of building technological competences, the feasibility study for the application is carried out in a relatively wide range of horizon, and later on, gradually focuses on effective applications by accumulating technological capabilities. That is, competence building is strongly related to the firm's past accomplishments. The notions of path dependency and cumulativeness have a strong foundation. Competence building centers in key areas to enhance a firm's core capabilities. As they gradually learn and accumulate their knowledge bases, firms are able to narrow the search process to find fruitful applications (Miyazaki, 1994).

Organizational management for dynamic environment

In order to improve management for innovative products, Osawa and Miyazaki (2006) developed an evaluation method for R&D project performance (Osawa and Miyazaki, 2006). They studied 17 projects of new product development from materials, parts, devices, and systems along the value chain. They classified major factors for failure as technical problems and market-related problems by quantitative analysis. They showed that selecting projects based on the evaluation method and management with allocating R&D resources is effective as a solution to overcome the valley of death for technology commercialization.

In addition, Teece put forward that strategic business management and dynamic change of organization are important for promoting innovations. That is, in order to adapt an organization to changing business environment, "dynamic capability" should be provided with the organization (Teece, 2007). Especially in the phase of exploring application and the business, Teece pointed out that the development and exercise of dynamic capability lies at the core of enterprise success, from his view point of organization management.

Among these, as for seizing opportunities, he stated that it is necessary to maintain and improve technological competence or complementary assets. And when an occasion is ripe, it is also important to invest in the specific technologies or design which is the most likely to be accepted by the market.

Research Questions and Hypothesis

When promoting new business, collaboration with lead users and flexible organizational management for dynamic environment are very important measures. In addition, trickle-up strategy for applying and diffusing new technologies works significantly.

However, in reality, business development faces various issues depending on the specific technologies or businesses and also depending on the phase of their businesses. Based on previous studies and considering actual businesses, the process is not always smooth sailing, and some questions will arise. For example, among the challenges that they are facing, what are the difficult challenges to overcome? In case they cannot be resolved, will the technologies which have been developed for specific products or businesses not be utilized?

So the following are listed as the research questions.

1. Difficulties in responding to market requirements
   - If the requirements of the users or market are satisfied, will the business develop?
   - If the requirements are difficult to be overcome by a single firm or within a short term, will the anticipated business, market, or the industry start to progress by promoting alliances with outside organizations such as collaborations with lead users?
2. Structuralizing deadlocks and difficulties to overcome
   - What are the complex deadlocks to overcome and is it possible to classify them?
   - When it is difficult to overcome or avoid the deadlocks directly, does the technology development have to be reduced or stopped, and does the commercialization have to be given up?

To consider expansion possibilities for other products or businesses beyond the initial target will be important for accumulating experience and learning. However, in many cases of new business development, it is very difficult to follow all the possibilities since only a few members are assigned for a project. Instead, it is important to select applicable businesses.

In order to promote the commercialization for assumed business, analyzing the deadlocks is important. Therefore, based on the results of the analysis, if it is difficult to resolve by a single firm or in the short term, exploring other applications will be promptly needed. In this study, classification of the deadlocks particularly in the initial phase of technology commercialization, and measures to avoid it are investigated by using the cases of mobile fuel cell and microreactor technologies.

Proposed Approach

Structuring of customer's requirements — "Shotgun approach" to customers

Figure 1 shows the outline of the structuring process of customers' requirements in the early stage. In this analytical framework, basic steps are expressed for technology commercialization as shown as "I. Beginning business", "II. Structuralizing and analyzing the market", and "III. Enhancement of R&D for effective market". In terms of the step on "II. Structuralizing and analyzing the market", project managers have to make an evaluation of the priority by classifying and analyzing "the customers", "required functions", "necessity technology", and "applied products and their market" which are shown in lower side of Fig. 1.

In general, engineers tend to accept full requirements specified by their customers. Despite the advantages of such an approach for accumulating knowledge and technologies for prospective applications, it is burdensome and resource intensive. Datar *et al.* (1996) point out that excessive customer input can create confusion and duplication of business development efforts, which ultimately increases time to market, through

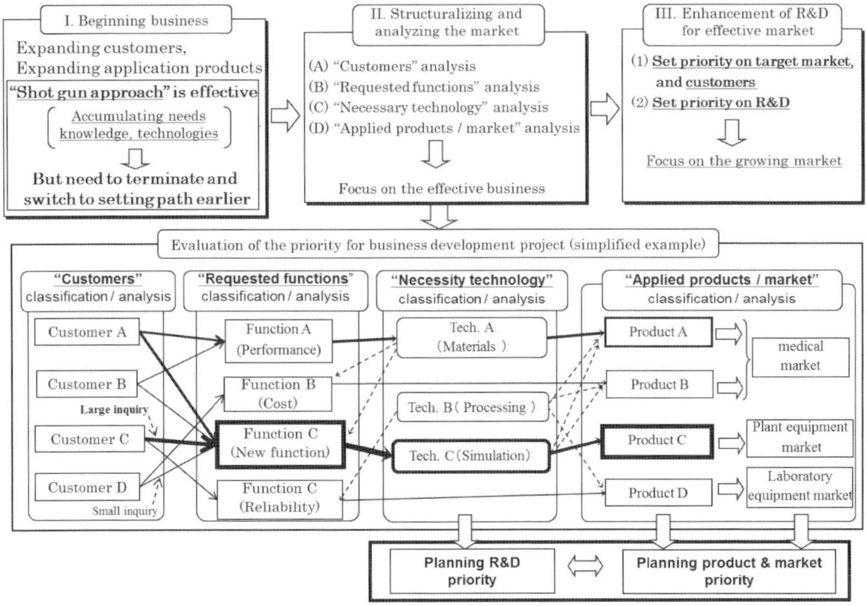

Figure 1: Structuring customer's requirements (analysis framework).

analyzing 220 new product developments in the electronic component industries (Datar *et al.*, 1996).

To identify and shape opportunities, enterprises must constantly scan, search, and explore across technologies and markets, both "local" and "distant" (Nelson and Winter, 1982).

For this reason, it is critical to structure and analyze the requirements from the market and customers, as well as set a priority on the targeted markets and customers. Knowledge from customers and markets will point out the appropriate directions for both R&D and business development. However, in case of structuralizing and analyzing customer's requirements, there are many cases which make it difficult to progress the business due to interacting inhibitory factors.

During the introductory phase especially in the formulation of the initial market and after the structuralization of those issues confronting the business, it is critical to set business paths which connect short-term and midterm business goals together considering the issues to be overcome.

Model for setting application path

Figure 2 shows the model for setting an application path. Collaborating with users from the beginning can be easy to confirm the direction of the development. Meanwhile, in case one needs to start from approaching prospective customers, a "shotgun approach" can be an effective method for exploring multiple possibilities. In this case, specifications such as performance, reliability will be required depending on the anticipated markets by each prospective customer.

In the introductory phase of commercialization, it is important to consider the following steps:

(i) "Shotgun approach" for searching multiple possibilities
Such an approach is effective for understanding and analyzing customer needs. However, it is necessary to terminate the shotgun approach and switch to setting path early, since a wide range of development causes fatigue for the business development team.

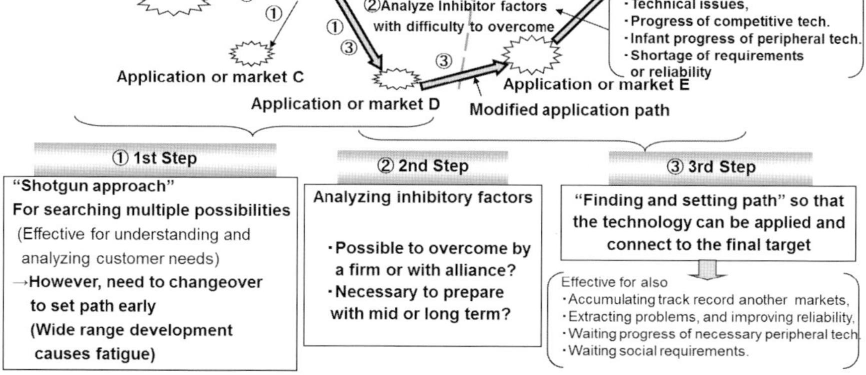

Figure 2: Model for setting application path.

(ii) Analyzing inhibitory factors
Difficulties and the period to overcome the issues have to be considered by a firm or through alliances by extracting and structuring the issues involved in each of the possibilities.

(iii) "Finding and setting path" can be applied to the technologies and connected to the final target.

This approach can also be effective for the following:

- Accumulating practical track records in other markets
- Extracting problems and improving reliability
- Waiting for the progress of necessary complementary technologies
- Waiting for the increasing level of social requirements

In the following section, an analysis of business progress is examined in the case of the DMFC for small mobile devices.

The Case of DMFC

Trend of Main Players

DMFC is a subcategory of fuel cell which uses methanol as the fuel. The main advantages are "small size", "no need to recharge", and "easy transportation of methanol" because of the energy density of the fuel. Therefore, the targeted markets are especially mobile equipments which require energy and density value to be more important than efficiency.

After 2003, Toshiba and Hitachi announced the development of a "Prototype" of mobile fuel cell for note PCs, cell phone, and fuel supply cartridge. In the market of mobile phone which is expected to be the largest market for mobile fuel cells, NTT DoCoMo and KDDI announced their intentions to develop fuel cells as the next generation power supply source for mobile phone in 2006. In an attempt to satisfy the goals of these telecommunication carriers, both Toshiba and Hitachi began to collaborate on developing the mobile fuel cells. In the same year 2006, Samsung also released an announcement on the prototype of personal digital assistant (PDA) and mobile battery charger for mobile equipment using DMFC. Furthermore, Matsushita Battery announced DMFC-prototype for mobile PCs in 2006.

Toshiba continued with the development of fuel cells for portable music players and notebook PCs. Moreover, in January 2008, Toshiba made public their intention to commercialize DMFC equipped products for their own mobile music player or note PC in 2008. Meanwhile, Hitachi was the only company engaged in the development and manufacture of DMFC for PDA equipment, prior to its mass production during Aichi EXPO 2005 for half year operation. Besides, Hitachi announced in 2008 the development of 100W portable electric source.

But as of 2016 it has not been commercialized yet, except for limited trial sale of 200 sets of portable battery charger by Toshiba at the end of 2009.

Classification of the Factors Affecting Commercialization of DMFC and the Current Situation

Currently, the performance of DMFC is badly affected by some technical issues such as "methanol crossover" in membranes. Manufacturers are trying to decrease the crossover, for example, by improving molecular structure of the membrane or by diluting methanol concentration; however, commercialization does not progress by only solving technical issues.

Table 1 shows the factors affecting commercialization of DMFC and a description of the current situation based on interviews at DMFC manufacture, news releases, and conference presentations.

Table 1: Factors affecting commercialization of DMFC and an analysis of the current situation.

Calssification of Factors		Current Situation
Issues by supplier side ↑	(1) Performance	• **Ready to produce. Technoloalcally completed including manufacturing deveolpment**. • **As the energy density of DMFC is twice as hiah as Li-ion battery, the competitiveness will grow with development progress**.
	(2) Cost	• Need to **keep cost down**, including assembling stack, separator.
	(3) Safety, Reliability	• Keep on improving safety issues of flammable full **especially for mobile phono due to "body touch usage" (Shortage of safety evdence by practical use)** • Improving environmental resistance supposing various situations.
	(4) Usability	• "**No time for electric charging**" is the most marked value. • If can be hybrid coexistence with Li-ion battery rather than a replacement • Need to improve for fuel consumption, heat and vapor generating during the work
↓ **Issues by market side**	(5) Size, Configuration	• Downsizing and slim consumption are desired
	(6) Standardization, Regulation	• Permission to bring fuel into airplane was realized (Relaxation of regulations) • Standardizations such as fuel cartridge connector do not affect on business start up • Need to improve sales network for fuel cartridges
	(7) User needs	• **Shortage of "necessity application" of DMFC, (Li-ion is enough)**

It can be said that most of the technical issues for production have been overcome. Ministry of Economy Trade and Industry (METI) of Japan also announced that the support for technological efforts on the development of DMFC has been completed, and it is already at the onset phase of commercialization.

According to the interviews from DMFC manufacturers, "cost down" and "high performance" are still issues for continuous improvement, but those technical issues such as improving the performance are not critical to start DMFC production. They are "ready to produce" by preparing production line.

Figure 3 shows an analysis of the inhibitory factors according to their position in the value chain. As can be seen, upstream actors such as "material" or "device" manufacturers regard "performance" and "manufacturing cost" as the major issues to deal with during the realization of the business. However, more critical issues such as "safety, reliability" and "lack of user

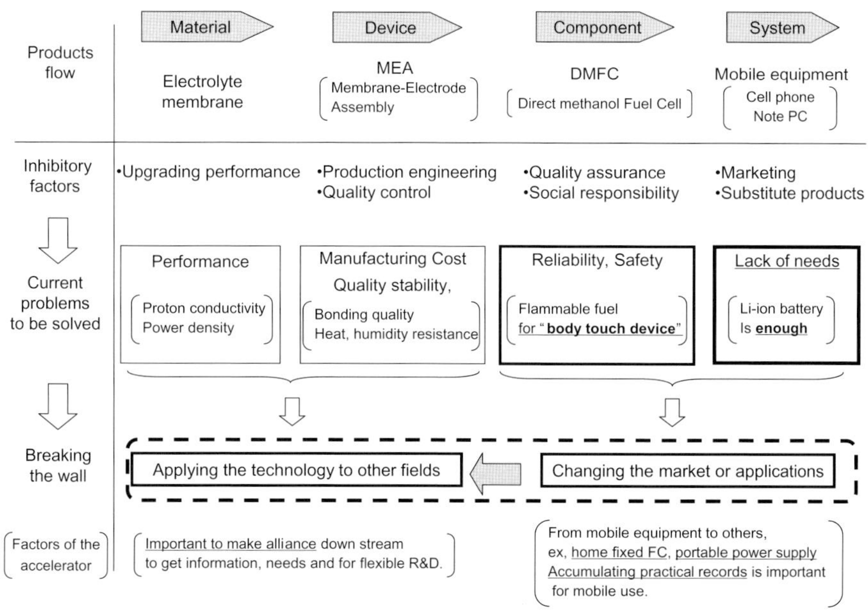

Figure 3: Analysis of the inhibitory factors and the factors for acceleration.

needs" are located downstream. Most of the DMFC manufacturers are large firms which are working on reliability issues very carefully. Especially for the use of DMFC as a power source for cell phones, the safety of the fuel cell must be sufficiently established since the device with flammable fuel is used in close contact with human body. In this situation, accumulating the "actual and practical use in the markets" will be necessary to overcome the safety and reliability issues by using current technology which has already been studied considerably. On the other hand, cell phone and notebook PC users regard the performance of Li-ion batteries as sufficient based on their current patterns of use. Therefore, in the early stages of commercialization, it is expected that DMFC technologies will be applied in other products since it takes medium or long term to solve these issues.

Firms developing small fuel cell technologies have explored their applications mainly for mobile equipments because their most remarkable characteristic is "no need for recharge". Table 2 shows the examples of the case for a small fuel cell. While, as mentioned earlier, the first product to market must be carefully selected due to a lack of track records, it will be possible to decrease the hurdles for commercialization even for mobile equipment such as portable power supply if it keeps certain distance from the body. In addition, further potential applications are power sources in electric wheelchairs, forklifts, and golf carts. The latter applications exploit the value that enables continuous usage without the use of commercial power source.

Example of Setting Application Path for DMFC

In the initial phase of the business, R&D direction is established by setting a target of their supposed applications. However, the first application will not always be released to the market.

Figure 4 shows an example of setting the application path for DMFC. At first, the direction of R&D is targeted to mobile

Table 2: Applications examples of small fuel cell.

Fields	Classification	Exampels of Application	
(1) Consumer use	(1) Mobile equipments	Cell phone, Note PC, PDA, Music Player, Camera, Video camera	← Initial Target ↓ Final Target
	(2) Outdoor power supply	Portable power supply, portable generator, Backup power supply, Power supply for leisure.	Candidate applications on the path
	(3) Home use fixed power supply	Home use fuel cell system	
	(4) Home appliance	Wireless electric machine tools, Lighting equipment, Eelctric torch	
(2) Industrial use	(1) Transportation	Automobile, Bus, Motorcycle, Golf cart, Forklift	
	(2) Medical equipment	Electric wheel chair, Medical power supply	

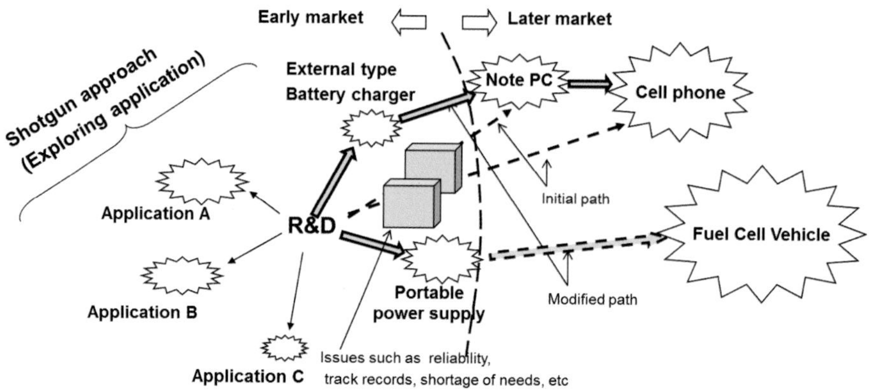

Figure 4: Example of setting an application path for DMFC.

phone and notebook PC as the market is expected to be large. But as mentioned, in the case of delay for the emerging market to grow due to inhibitory factors, the exploration of alternative markets of applications should be considered, such as electric screwdriver, video cameras for professional use, and electric wheelchairs, even though the size of those markets are not so large compared to the initial targeted markets.

Modified application paths should be defined by considering the easiness of their technical deployment, or their accumulating reliability.

The most important point in gaining the competitive position in emerging markets is the setting of appropriate paths which provides opportunities for accumulating knowledge and technologies to target the final markets.

Therefore, during the introductory phase of a business development, it is important to set a path for the final target markets after surveying multiple possibilities. The modified application path will have a possibility to create cash flow, and result in accumulating knowledge, technological competence, practical reliability, and anticipating social requirements with increasing competency.

Case of Fluid MEMS

Outline and History of the Microreactor Technology

Microelectromechanical systems (MEMS) devices which are made by the application of microfabrication technologies have been applied to various fields such as sensors, optics, and electric power devices among others. The research and development for the microreactor has been mainly conducted in German and Japanese institutes since the late 1990s as a fine control technology for chemical "blending", "reaction", "condensation", "extraction", and "separation" by forming microfluid channels which ranges several micrometers to several hundreds of micrometers in width. In Germany, the Karlsruhe Institute of Technology and the Institute for Microtechnology Mainz (IMM) are active in the development of basic technologies. In Japan, the university–industry collaboration among Kyoto University, Tokyo University, Tokyo Institute of Technology, Japan Chemical Innovation Institute (JCII), and New Energy and Industrial Technology Development Organization (NEDO) has made substantial progress up to now.

In 1997, when microreactor technology was introduced to Japan, the first research output was reported by JCII, named "Research on the microreactor technology for chemical synthesis". Moreover, the project "Micro analysis, production system project" — supported by NEDO from 2002 to 2005, stressed research by university–industry collaboration. Through applying microreactor technologies, it may be possible to replace conventional batch production system in large scale chemical plants with the continuous production system through desktop sized equipment. For some industries such as chemical, pharmaceutical, food, and beverages, there are high expectations on the microreactor technology by creating or discovering materials and R&D process on fine chemicals (electronic materials, pharmaceutical intermediary body, etc.) to mass production of industrial materials. The technology may achieve low cost production, saving spaces, and material resources by the precise control of the reaction processes. Furthermore, efficient energy consumption may also be expected since the microreactor does not require large scale cooling facilities compared to the conventional manufacturing plant which make use of exothermal reactions. In addition, lead time of the process development for mass production can be shortened by "numbering up" which simply consists of the parallel alignment of the microreactors, instead of "scale up process" which is the conventional step from test-tube, beaker, flask-work to bucket, barrel, and plant (Fig. 5).

Issues for Commercialization

Application diversity

Various issues have to be resolved when the microreactor technology is applied in manufacturing, since the candidate processes of substitutable reactions cover a broad range, as well as synthesized compounds are almost countless. For this

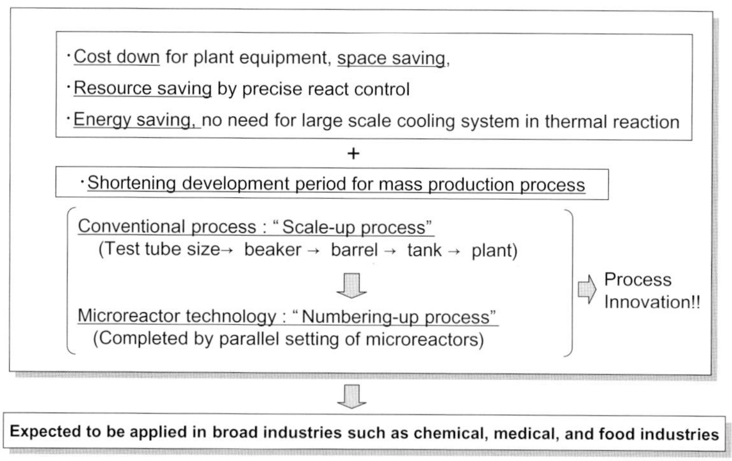

Figure 5: Effects and values of microreactor technology.

reason, development load for practical realization increases with expanding candidate applications. For example, in the field of organic synthesis, the reaction processes include various syntheses such as liquid phase intermediate synthesis, polymerization reaction, emulsification, catalytic synthesis reactions, particle synthesis, and their combinations. In order to apply the microreactors, the processes of generating variety of chemical compounds have to be investigated in each reaction, respectively.

In considering the applicability of each manufacturing process, "customizing the technology" must be required, such as optimizing parameters or operation methods by studying the mechanisms of those chemical reactions in such a "tiny space". In other words, the specific issues for microreactors have to be resolved, for instance, setting parameters to improve its yield among others or to avoid channel blockage by generated small materials. In this case, those issues may be peculiar and unforeseeable for the researchers in the field of microreactor technologies.

Technology diffusion

Here, regarding the microreactor as a potential industry, "diffusion of laboratory system in early stage" has important implications before the diffusion of industrial microreactor applications, because the laboratory system makes it easy to survey basic parameters for preparing the supposed manufacturing processes.

As already mentioned, the basic technology development has been studied in universities and public research institutions. After collaboration with "public" research institutes, user companies need some steps to transfer the technologies, and to study or to apply to their own individual applications in each company. In the following section, a successful case study of a company which succeeded in this process is described. Furthermore, in order to accelerate the diffusion of microreactor technology, a number of "potential candidates of user companies" which are interested in the emerging technology, hopefully should set up an environment for promoting their own R&D individually. From this perspective, spread of the laboratory system can be regarded as an important step as a precursor of microreactor diffusion.

In resolving these issues, CPC-Cellular Process Chemistry Systems GmbH, a German specialty firm founded in 1999, is focusing on development of the laboratory systems based on microreactor technologies. In addition, they have been conducting collaborative research with user companies in the experimental phases, and more than 20 cases of the successful synthesis are shown in their web page. However, only a few cases have reached the commercialization stage, since other unsolved issues still remain such as the development of equipment for continuous pumping of liquid solutions, which has to be optimized for each reactant.

However, for the user companies, "introducing a laboratory system, i.e., installing the experimental equipment" is not the final goal, but only "efficient commercial production" is the final goal. Hence, one has to also take into account setting up the manufacturing plant. That is, most of the manufacturers which produce materials, so far, design and develop their production processes, facilities, and lines by themselves depending on the materials.

Therefore, in terms of creating a new industry, players who are trying to deploy the microreactor business, need to do consulting work such as understanding users' requirements, customizing microreactors technologies, and proposing solutions. So, the target of commercialization should be placed on both "the laboratory system" for introducing the art to the candidate users and "the plant system" for commercial production.

As for the diffusion of the microreactor technologies, "increasing the load of the support for applying to each process" will be a major issue, in addition to "the application diversity", even after focusing on specific applications. For these issues related to the application diversity and the development load, "providing a laboratory system" can contribute to resolve them since it is possible to optimize parameters "by the user's own experimental examinations". In other words, by providing tools such as the laboratory system, one can absorb and exploit the user's development capabilities. It can also be an important factor to accelerate the business and its commercialization. Furthermore, since the user company also tries to promote the application of the technology to their manufacturing process based on the knowledge from microreactor business player, "to absorb and to exploit partner's abilities mutually" is a significant concept (Fig. 6).

Customary cultural habits

On the other hand, when the user companies enter into a stage of process development, engineers also usually face unforeseen situations, which cannot be forecasted by the traditional preliminary experimental procedures using *test-tube* or *flask-works*. The burden of the engineers is not light since they have not experienced the development process. Most of the user companies are interested in microreactor technologies, however, they also have to face and overcome these *traditional cultural habit issues such as changing conventional process development like flask-work*. The microreactor technologies could offer advantages if they apply it from fundamental phases for synthesis

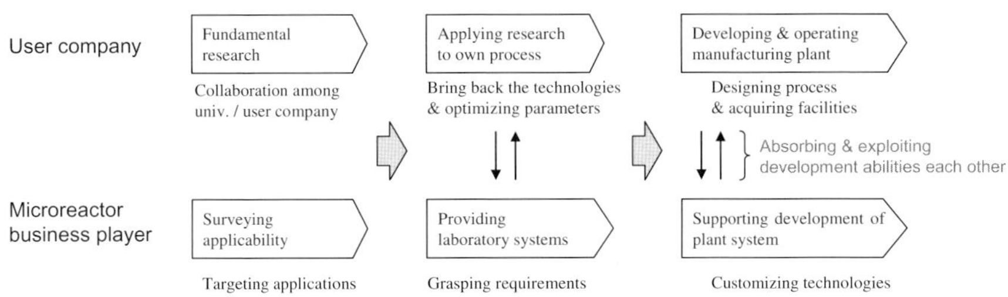

Figure 6: Diffusion process of the microreactor technology.

experiments because "the numbering up" can be replaced by "the scale up" for an effective process development. Thus, the engineers need to change their mind and the traditional style even for basic experiments when introducing the new process innovation such the microreactor technologies.

Nelson *et al.* had conducted research on organizational issues for promoting a business. In their study, they pointed out that a bias works for a new approach or method which differs from conventional efforts (Nelson and Winter, 1982).

If there are standard and routine methods which have traditionally worked, resistance may tend to appear for new approaches toward new business development (Teece, 2007). In case of promoting new business development, organizational issues may arise even within firms. According to Nelson and Winter, existing corporations tend to avoid radical innovations which destroy competencies, but prefer increasing competency with incremental improvements (Nelson and Winter, 1982).

In a similar way, for instance, in the introductory phase of 3-dimensional Computer Aided Design (3D-CAD) as a drawing design tool for manufacturers, there were some instances that engineers who had traditionally been using 2D-CAD resisted to switch the drafting process.

Around the late 1990s to the early 2000s during early diffusion stage of 3D-CAD, some bias to introducing new technology by engineers were heard; for example, "We can design by a conventional method or traditional way without any problem", "I do not feel the necessity of 3D-CAD specifically", or "I use 2D-CAD, because we can ask operators to convert 3D-CAD data when we need Computer Aided Engineering (CAE) like structural analysis or fluid analysis". These comments were heard not only from the author's own experience as an engineer in mechanical electrical engineering but also were heard through interviews from engineers in the construction industry, plastic parts industry, plant engineering industry, and so forth.

In order to overcome such issues and absorb or apply new technologies by accepting considerable load for learning new techniques, it is necessary to understand the advantages of the new technologies by their organizations ranging from staff engineers to managers.

Motivations for shifting to 3D-CAD may be shown by the following examples:

- By drawing a 3D design from the beginning, the conversion process from 2D to 3D can be omitted when running the CAE analysis, so the process can be expected to shorten the development time.
- It will be possible to make prior evaluation with various simulation methods by passing and sharing data from parts to final products.

In other words, from design to production, or from parts design engineers to final product users, it will be possible to accept the effectiveness through the development process.

Back to the microreactor and considering the situation to introduce a new process, in addition to some hurdles to learn new techniques, a psychological feeling of resistance may appear, since they need to change the basic familiar experimental approach such as using a test tube. "To change a process itself for product development or process development for manufacturing" can be regarded as a cultural and habitual change. Therefore, "a bias for inexperienced process" should be removed. To do this, improvements, advantages, or possibilities caused by the technology should be showed and understood by engineers.

Successful Case of Commercialization

At least, clarifying the advantages has to be shown when a technology is applied to practical commercialization.

The advantages of microreactor are:

(i) Low cost facilities (equipment, maintenance, and operation cost)
(ii) Space efficiency (desktop size production instead of large scale batch plant)
(iii) Materials resource saving (reduction of material usage by improving yield)
(iv) Energy saving (large reduction of energy consumption for cooling down the reaction heat)
(v) Automation and work load saving (reduction of manpower by automatic operation)

Appropriate reactions using the microreactor are:

(i) Exothermal reaction with high temperature
(ii) Rapid reaction

In 2007, Nippoh Chemicals which joined the NEDO project called "Micro chemical process technology research association" from 2002 to 2006, started commercial production of "Trimethyl orthoacetate" by using a microreactor.

The project manager proposed and highlighted the need to replace the old facilities with the microreactor, since the based equipment requires periodical inspection and repair. He succeeded in persuading the internal decision makers by insisting on the inefficiencies in periodical maintenance cost. The manager then listed up the candidate reactions that can be produced by using the microreactor in the company, and thoroughly researched its application into various productions. On this occasion, the lead manager conducted feasibility studies and concluded that benefits of reducing cost by several million yen would arise compared to the maintenance cost of old facilities fluctuating around tens of million yen to several hundred million yen.

Furthermore, by introducing the microreactor produced 1000 ton of Trimethyl orthoacetate since the reaction time was

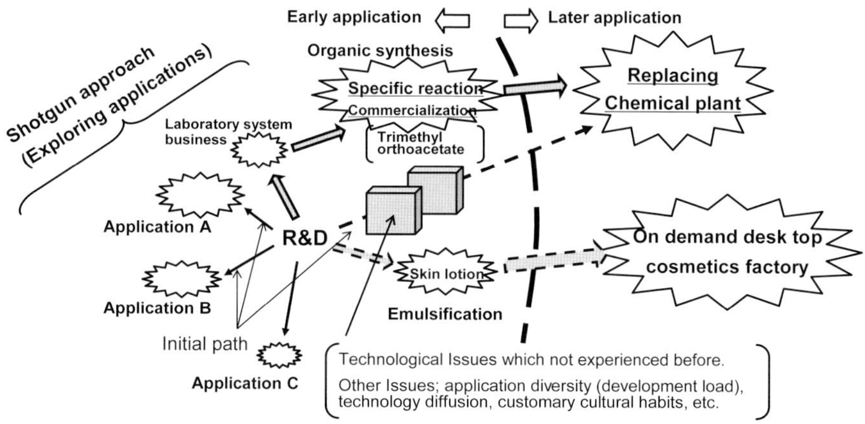

Figure 7: Example of setting application path for microreactor.

shortened from 30 h to 200 s. The experiences of Nippoh Chemicals with microreactor technologies allowed the firm to overcome the enormous belief that microreactors are not capable of reaching capacities in the mass-production range. Nippoh Chemicals decided to expand the use of microreactor technologies to other materials apart from Trimethyl orthoacetate.

Conclusions

In the case of developing a new business based on emerging technologies that do not exist so far, the anticipated applications or markets do not emerge as expected.

In this study it was found that structuring and analyzing inhibitory factors are significant issues, in addition to extracting user requirements and their realization.

As mentioned, the deadlocks of new technology commercialization are not only requirements of improving performance or decreasing cost, but also further progress of conventional competitive technologies, infant progress of necessary peripheral technologies. Furthermore, this study showed additional essential factors of deadlocks ranging from "primitive and social demand of the technology", to "customary cultural habits".

Particularly in the introductory phase of commercialization, it is important to consider the following steps for accumulating technologies, experience, and acquiring competitiveness:

(i) Shotgun approach for multiple possibilities (however, need to changeover)
(ii) Extracting and structuring the issues from each possibility
(iii) Setting application path to final target market or applications with accumulating knowledge and technologies

Even though the cases of DMFC is product innovation and the microreactor is process innovation, their candidate applications cover a broad field in common, so that it has to be surveyed by analyzing and overcoming deadlocks for their successful commercialization. Therefore, establishing and modeling "paths to create a final market" is crucial not only for commercialization but also for accumulating knowledge, building technology competence, and anticipating social requirements with increasing reliability.

In addition, there are some issues which are difficult to solve by a firm by itself. For example, for microreactor, to involve or to exploit users' developing ability will be an effective acceleration factor for technology commercialization by providing laboratory systems. It is expected that microreactor technology will have a wide range of applications. However, there are still serious issues in the commercialization of these technologies such as technological issues, development load issues, as well as feasibility issues. Furthermore, the challenges that engineers face and have to overcome are sometimes technological issues which they have never experienced before. Therefore, alliances with external partners and bringing technologies from different fields are important. This successful example is the result of careful analysis with regard to such issues, as well as execution of a correct strategic management. Through the successful example presented in this section, the recommended strategy is to begin with those technological applications expected to bring the highest economic advantages. Here, the technologies are assimilated, knowledge is accumulated, and it is then possible to explore new fields of applications.

Microreactor is a new technology that changes the process itself for creating and manufacturing materials. To introduce the technology can be considered as a process innovation, that is, it has a higher potential to revolutionize the industry structure. Exploring applications is mandatory for business development of infant technologies, however, their commercialization and diffusion are not easy. Some trial cases at an experimental level have been seen in some institutes so far, but a manufacturing trial for the commercialization has been currently challenging to the existing mature technologies. It is necessary to set priority by examining the commercialization with a broad perspective such as technical difficulties of applying them to supposed reaction, reduction effect of maintenance costs and running costs, without bias to the particular reactions or applications.

In this study, a new concept that leads to the goals with avoiding deadlock factors for commercialization has been highlighted. That is, in order to overcome issues for commercializing emerging technologies which do not exist so far, a new concept of "setting technology application path binding immediate goals" with actively accumulating knowledges and experiences will lead to desired final target.

The development of new business does not allow arguments such as "*Business will not progress without the proper conditions*". In contrast, our argument made throughout this chapter is that project managers should manage so as to "*Progress the business through creating a market or realizing applications within the current conditions*".

References

Chesbrough, H. (2003). *Open Innovation*. Boston, MA: Harvard Business School Press.

Datar, S., Jordan, C., Kekre, S., Rajiv, S., and Srinivasan, K. (1996). New product development structures: The effect of customer overload on post-concept time to market. *Journal of Product Innovation Management*, Vol. **13**: 325–333, Elsevier.

Day, G. (2000). Assessing future markets for new technologies. In Day, G., Schoemaker, P., & Gunther, R. (Eds.), *Wharton on Managing Emerging Technologies*. New York, NY: John Wiley & Sons.

Drucker, P.F. (1985). *Innovation and Entrepreneurship*. New York, NY: Harper & Row.

Freeman, C. and Soete, L. (1974). *The Economics of Industrial Innovation*. Cambridge, MA: MIT Press.

Handfield, R.B., Ragatz, G.L., Peterson, K.J., and Monczka, R.M. (1999). Involving suppliers in new product development. *California Management Review*, Vol. **42**, No. 1: 59–82.

Lieberman, M. and Montgomery, D. (1988). First-mover advantages. *Strategic Management Journal*, Vol. **9** summer: 41–58.

Miyazaki, K. (1994). Search, learning and accumulation of technical competences; the case of optoelectronics. In *Industrial and Corporate Change*, Vol. **3**, No. 3. New York: Oxford University Press.

Nelson, R.R. and Winter S.G. (1982). *An Evolutionary Theory of Economic Change*. Cambridge, MA: Harvard University Press.

Osawa, Y. and Miyazaki, K. (2006). An empirical analysis of the valley of death: large-scale R&D project performance in Japanese diversified company. *Asian Journal of Technology and Innovation*, Vol. **14**, No. 2: 93–116.

Schumpeter, J A. (1934). *The Theory of Economic Development*. New Brunswick, NJ: Harvard Economics Studies v.46.

Teece, D. (2007). Explicating dynamic capabilities: The nature and microfoundations of (sustainable) enterprise performance. *Strategic Management Journal*, Vol. **28**: 1319–1350.

Utterback, J.M. (1994). *Mastering the Dynamics of Innovation*. Boston, MA: Harvard Business School Press.

Veldhuizen, E., Hultink, E.J., and Griffin, A. (2006). Modeling market information processing in new product development: An empirical analysis. *Journal of Engineering Technology Management*, Vol. **23**: 353–37.

von Hippel, E. (1988). *The Source of Innovation*. New York, NY: Oxford University Press.

Chapter 14

Technology Diffusion: Computational Fluid Dynamics (CFD) for Flow Simulation in Greenhouses

L. Pretorius, S. Kruger† & S. J. Benade‡*

Technology can be considered as the stock or inventory of knowledge, competencies and capacities that a firm has to compete effectively in the market place (Dorf and Byers, 2008; Nieto, 2004). Technology therefore forms part of the technology innovation process. In this sense, innovation (management) is more of a rate dependent entity whereas technology is a level or stock quantity. Technology is both an input as well as an output of the innovation process (Nieto, 2004). In this chapter, the essential body of knowledge of computational fluid dynamics (CFD) will be considered as a technology. It will also be illustrated in some sense how CFD can be used in the innovation process. CFD is sometimes also considered as a useful tool in the innovation process.

Drejer (1997), for example, takes a tool, system, and value view on technology. The system view implies for instance that a user will be influenced by the technology. The value perspective

*Department of Engineering and Technology Management, Graduate School of Technology Management, University of Pretoria, South Africa.
†Department of Mechanical Engineering Science, University of Johannesburg, South Africa.
‡Department of Engineering and Technology Management, Graduate School of Technology Management, University of Pretoria, South Africa.

of technology can on the other hand imply usefulness of a technology. In this chapter, the usefulness of CFD as a technology in the hands of the designer or innovator is the focus. The designer as part of the engineering team is then instrumental in the usefulness of CFD as technology. It is then sometimes also useful to be able to compare the performance of technologies in some sense. In this chapter, the growth performance of different CFD technologies will be compared.

Drejer (1997) quotes Badaway that "management of technology links engineering, science, and management disciplines to plan, develop, and implement technological capabilities to shape and accomplish the strategic and operational objectives of an organisation." This definition of technology management places for instance research technology and product development in direct relation to the marketing and human resources functions in the firm. It is also implied that innovation can be considered a driver in the process of management of technology. The rate of change of CFD as technology or the innovation of CFD will be the focus in this chapter. The rate of innovation in CFD typically has an effect on, for example, the technical skills required in design teams of organization responsible for product development.

Fluid dynamics and heat transfer are both natural phenomena that impact heavily on the engineering development and fundamental operation of many new product developments such as greenhouses for agricultural applications (Kruger and Pretorius, 2007), photovoltaic solar water pumps (Van der Merwe, 2002), dry cooled condensers (Van Staden *et al.*, 1998), the ink-jet printer (Clymer and Asaba, 2008), and so forth.

The innovation process of such new thermal equipment is directly influenced by the capability to do effective thermodynamic and fluid dynamic design. The Navier Stokes equations are generally utilized to model such physical phenomena. These equations are generally nonlinear partial differential equations and analytic solutions are not easy to obtain (Incropera and De Witt, 2006; Versteeg and Malalasekera, 2007).

These equations can be discretized and form the basis of CFD to obtain approximate solutions to more general flow and heat transfer situations (Versteeg and Malalasekera, 2007). In this vein, Sturgess (1996) also points to the development use and role of CFD in the design of enabling technologies such as the fuel pump in gas turbine combustors.

In the light of the aforementioned CFD as a body of knowledge used to effect a competitive position for a firm may be considered to be a technology and managed as a technology as part of the innovation process. An explanation of how, for example, CFD as body of knowledge and as part of the design activity was used in the competitive positioning of the ink-jet printer is provided by Clymer and Asaba (2008). This design activity role of CFD is also emphasized in this chapter as part of the discussion of CFD in greenhouses.

One way of measuring or characterizing the growth of technologies is through the use of bibliometrics. It is mentioned by Bae *et al.* (2007) that bibliometric analysis can be usefully employed to gauge technology maturity. It is expected that in a technology growth phase the number of academic articles in the specific field of technology should also increase. This forms the basis of bibliometric analysis. The results of these analyses may also be effective in assessing technology trends (Bae *et al.*, 2007; Nieto, 2004). As an example of the use of bibliometrics, Daim *et al.* (2006) uses this technique to assist in the forecasting of emerging technologies such as fuel cells.

The application of Scientometrics to assess the extent of nanoscale research in South Africa is furthermore illustrated by Pouris (2007). He uses this field of study related to bibliometrics to discuss trends and growth in nanoscale research in South Africa (SA) in relation to the international context. The study is intended to influence national policy of SA on nanoscale technology.

The process of adoption and growth of technology and products by a potential market is under certain circumstances described by a diffusion model. One of the earlier models used for this diffusion process is the Bass diffusion model (Bass, 2004).

A combination of bibliometrics and the earlier Bass diffusion model is used in this chapter to show trends in CFD technology development that can be used by executive management to assess their approach to technology management. This is also illustrated for the case of CFD simulation of the internal flow environment of an eight span greenhouse.

This chapter is an extension and contextualization of work presented in conference and journal papers at ICMIT08 and ASME08 by the authors (Kruger and Pretorius, 2008; Pretorius *et al.*, 2008, 2011).

This chapter is organized as follows: In the next section, the methodology followed during this research is presented. The methodology is presented in a fashion that integrates the supporting and relevant literature review on bibliometrics diffusion models and CFD technology. The next section introduces the implementation of the Bass diffusion model in a spreadsheet simulation environment that is conducive to estimate diffusion characteristics of a technology from a set of real data. The technology data simulated includes technology trends for CFD. This is followed by a discussion and comparison of simulated CFD technology diffusion S curves and possible usefulness in forecasting. In the conclusion section, some issues for further research resulting from the technology diffusion simulation approach taken in this chapter are discussed. This also includes some reference to the usefulness of system dynamics in this technology domain.

Methodology and Literature Review

In this section, the methods used in this chapter to measure and compare appropriate information pertaining to CFD as a construct are addressed. Furthermore, the method used to model diffusion of CFD technology and the basic essentials of CFD as a technology are also briefly described.

Bibliometric Analysis

It is mentioned by Bae *et al.* (2007) that bibliometrics is essentially a process of measuring text and information. In this chapter, bibliometrics is used for measuring constructs relating to CFD. Bibliometrics can also be used to appreciate and compare some events that occurred in the past or even attempt to make some technology forecasts. This is also illustrated in some sense by Daim *et al.* (2006) in their research on emerging technologies.

Jordan (2007) specifically mentions the use of bibliometrics to trace technological change in his study of laptop battery future. The usefulness of this technique is illustrated by Jordan when he uncovers and compares typical S-curves for Li-ion and lead acid batteries that indicate saturation for these technologies from the databases. He interprets this as a breakthrough for a new technology may be impending.

In this chapter, the technology trend for CFD as a technology is traced via bibliometrics. Essentially, the number of times constructs pertaining to CFD occur in certain periods are measured and evaluated. The main data source used in this chapter for these CFD construct measurements is Google Scholar (2008). This may be limiting on the research illustrated in this chapter but will be addressed in future research.

Bass Diffusion Model

The discrete equivalent of the basic Bass diffusion model is described by the following equation (Bass, 2004; Bass curves, 2008; Mahajan *et al.*, 1995):

$$N_t = N_{t-1} + p(m - N_{t-1}) + q\frac{N_{t-1}}{m}(m - N_{t-1}) \tag{1}$$

The symbols in Eq. (1) have the following meaning:

N_t: number at time t

p: coefficient of innovation or external influence

q: coefficient of imitation or internal influence
m: total market size

Further versions of the generalized Bass Diffusion model are also available (Danaher *et al.*, 2001; Mahajan *et al.*, 1995). In these versions, the effect of substitution is also considered. In this chapter, the aim is first only to show that CFD as a technology also has some saturation characteristics and that the Bass diffusion model may be employed to assess and compare some of these technology trends. The effect of substitution and cross pollination between technologies and for example, second generations is not considered explicitly in the numerical model. The diffusion of CFD technology in greenhouse applications is also explored.

The coefficient of innovation (p) is typically indicative of the adopter's action to the product mainly because of the perceived new benefits whereas the coefficient of imitation (q) indicates a pressure on the adopter because the product seems to belong to a usual set that he has to have. The coefficients of innovation and imitation can be used to compare diffusion trends of technologies and this is illustrated for CFD as a technology in this chapter.

CFD Technology

In CFD a numerical solution of a set of partial differential equations, typically the Navier–Stokes equations are obtained (Incropera and De Witt, 2006; Versteeg and Malalasekera, 2007). These partial differential equations govern the transport of mass, momentum, and energy in moving fluids (Versteeg and Malalasekera, 2007). The laws governing the transport of the above quantities are: conservation of mass, Newton's second law of motion, and the first law of thermodynamics (Van Staden *et al.*, 1998). Finite volume discretization is the first step in solving these transport equations for a specific domain where, for example, a thermal design may be required.

In this method, the solution domain where a thermal or flow design may be required is subdivided into a finite number of small control volumes, that correspond to the cells of a computational grid. Discrete versions of the integral form of the continuum transport equations mentioned above are applied to each volume. The objective of this method is to obtain a set of approximate algebraic equations representing the continuum transport equations. An algebraic multigrid solver can be used to solve the resulting algebraic equations (StarCCM+, 2006). In the case of this chapter, the design focus is the flow and thermal domain of a greenhouse.

A more detailed description of a discretization procedure such as the one briefly described here can be found in Patankar (1980). In order to include buoyancy source terms in the momentum equation, a gravity model can be activated in the resulting CFD simulation software. To approximate the buoyancy effects in flow, the Boussinesq model has been implemented for the greenhouse design case illustration presented in this chapter by selecting a constant density flow.

Model and Results

In this section, the results achieved from implementing the basic Bass diffusion model, represented in numerically discretized form by Eq. (1), in an Excel spreadsheet simulation environment are presented and briefly discussed. The basic process of determining the best model parameters for a test case of personal computer (PC) diffusion into a market is firstly described followed by a discussion of determining the best model parameters for a case of CFD technology diffusion from bibliometric data. It should be noted that in this section the capital case parameters M, P and Q correspond to the small case parameters m, p, and q mentioned in Eq. (1) for the Bass diffusion model. Thereafter, the case of modeling CFD technology for greenhouse design is presented. Lastly, a comparison of the diffusion of two different CFD software codes used as technologies for design purposes is made.

CFD Technology Trends

To compare the currently developed Bass model by the authors, using Eq. (1) as a basis for a spreadsheet application, the data presented by Shang *et al.* (2004) for PC demand were fitted to a Bass curve by means of optimization of the Bass diffusion parameters m (total market size), p (coefficient of innovation), and q (coefficient of imitation). Data for PC demand correlate to a number of months reported in 1988.

This Bass model used is an original version of the model introduced by Bass (2004). It is adapted for numerical solution in a spreadsheet application. There are further versions of the Bass model available including for instance, the effect of marketing (Danaher *et al.*, 2001; Mahajan *et al.*, 1995).

The optimized parameters obtained by the author with Excel's solver (Gotfried, 2007) are shown in Table 1. It should be evident that the current model produces similar values to that suggested by Shang *et al.* (2004) for the specific PC market data considered. The simulated value of 1793×100 for eventual demand (eventual potential market segment size) compares well with the value of 1800×100 produced by Shang *et al.* (2004).

It is therefore assumed that the basic Bass diffusion model utilized in the current computer simulation will suffice for the intended purpose of at least qualitative exploratory forecasting and trend assessment. It is however also appreciated that some

Table 1: Optimized diffusion parameters for PC demand.

	M	P	Q
Current model	1793.028	0.035295	0.190435
Previous data []	1800.32	0.035	0.188

of the PC demand data supplied may have been infected with resale data which according to Shang *et al.* (2004) may corrupt the simulation. The trend is, however, considered to be in order for the current trend analysis purpose.

Having established the appropriateness of the current model for at least some available PC data as indicated in Fig. 1 some confidence exists in the model developed by the authors and it is deemed fit to proceed in applying the model to the case of CFD technology. PC as a hardware technology has in this section also been shown to have diffusion and saturation characteristics. Many software CFD codes as technologies are currently also available for use on certain PC's. It is thus deemed appropriate to examine and compare the diffusion characteristics of CFD technology.

The bibliometric data used to assess the Bass curve shown in Fig. 2 for the CFD technology trend are derived from research articles in Google Scholar (2008) with words, for example, "CFD Technology, year A to year B". In this process, the time of near zero publications is determined iteratively in Google Scholar. Thereafter, the number of appropriate articles is determined cumulatively from the start date for the applicable technology.

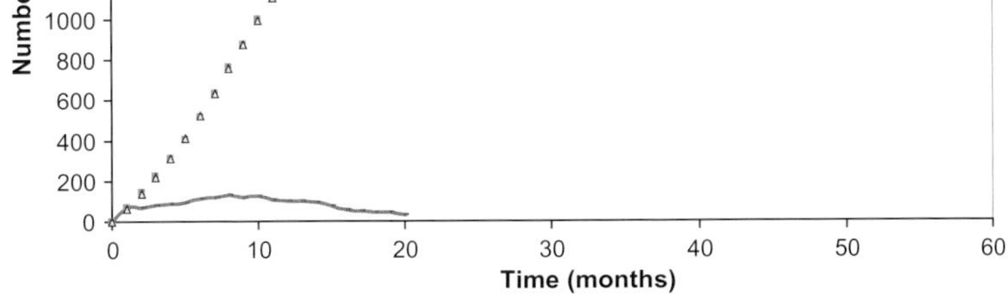

Figure 1: Demand trend for PC's compared with Shang *et al.* (2004).

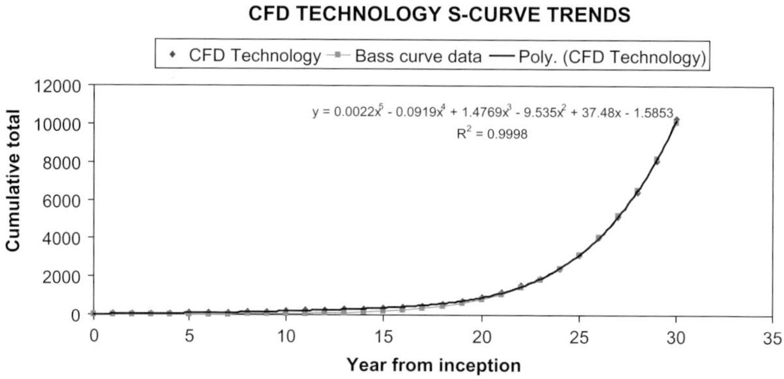

Figure 2: Technology trend for CFD.

The optimized Bass diffusion parameters for the CFD technology trend using the numerical form of Eq. (1) in collaboration with the bibliometric data for "CFD Technology" as obtained with the numerical Bass model in Excel's solver are shown in Table 2. It should be evident that the current model is in good agreement with data for the first 30 years of the CFD technology development. The CFD technology is still far from maturity as indicated by the relatively large value of 28,676 articles for *M*. The *Q* value of 0.322 is in the range suggested by previous literature (Bass, 2004; Shang *et al.*, 2004).

Year zero data correlate to 1976 data. What is also interesting to note is that bibliometrics may be used to establish in some sense even the inception of a technology such as CFD.

CFD Research Trends for Greenhouses

To illustrate the power of this CFD technology as design tool excerpts of CFD simulation results obtained for an eight span greenhouse by the authors (Kruger and Pretorius, 2008) are presented. Table 3 summarizes the design temperatures and velocities used for the wall boundaries of the greenhouse in the numerical CFD model, while Table 4 is a summary of the properties used in the k-epsilon turbulence flow model.

Table 2: Optimized diffusion parameters for CFD technology trend.

	M	P	Q
CFD technology	28676.69	3.59787E-05	0.322255

Table 3: Boundary conditions used in CFD model similar to Ould *et al.* (2006).

Parameters	Values
Inlet Air	
Velocity at 6 m [m/s]	**1.4**
Temperature [°C]	**22.2**
Temperature [°C]	
Outside Air	22.2
Outside Ground	27.9
Inside Ground	27.3
Roof	33.6
Plastic Central Partition	31.3
Glass Walls	29.1

Table 4: K-epsilon turbulence model parameters used.

Under relaxation factor	0.8
Convergence tolerance	0.1
Epsilon	0.0
Turbulent viscosity (under relaxation factor)	1.0

The power of CFD to simulate flow and temperature conditions inside the greenhouse for design purposes should be evident from Figs. 3 and 4. The existence of recirculation flow fields with inside velocities of 0.1–0.2 m/s due to the external flow as well as buoyancy inside is especially the first and second span is amply and visually illustrated using CFD as

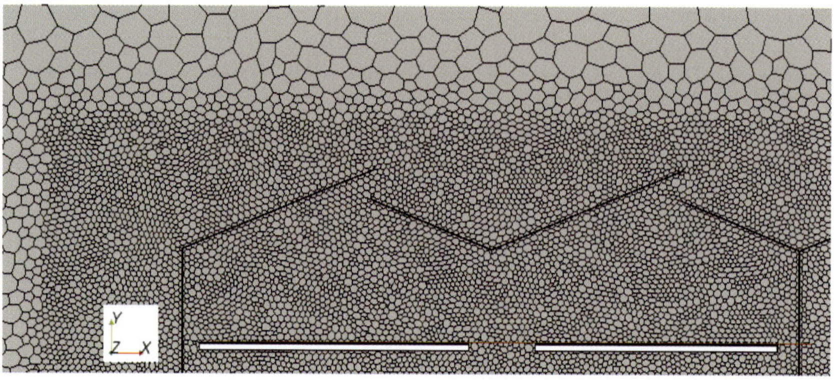

Figure 3: Refined polyhedral discretized mesh around the greenhouse.

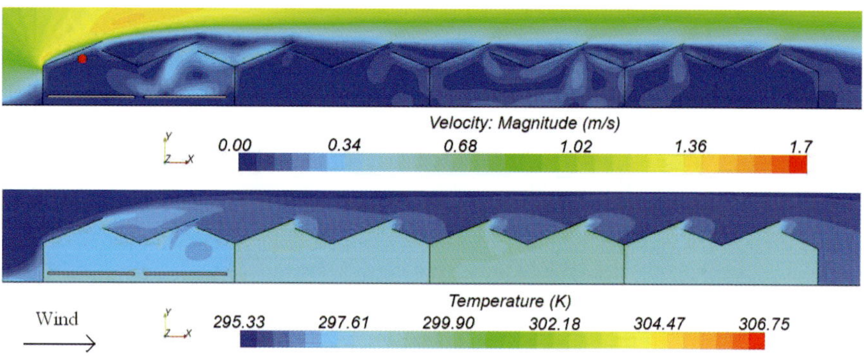

Figure 4: Velocity and temperature contour plots.

technology. The existence and extent of recirculation flow fields illustrated here using CFD may influence the eventual design of a greenhouse.

The question that arises now from a CFD design point of view is to what extent CFD as a design technology has diffused into the greenhouse application area. Delving further into the technology of CFD development this case of application of CFD to greenhouse environment simulation is assessed via bibliometric data for CFD in greenhouses from year 1998.

The bibliometric data used to simulate the Bass curves shown in Fig. 5 are derived from research articles in Google Scholar (2008) with words, for example, "CFD Greenhouse, year A to year B".

The optimized Bass curve simulated in Excel again using Eq. (1) for a fit of all the available bibliometric data to year 9 is shown in Fig. 5. If only bibliometric data up to year 8 are used to fit a Bass curve it can be shown that the prediction is 6.77% in error at time nine.

With the full set of data a saturation value *M* of 982 units is predicted with the Bass model. If only data up to time eight are used to predict *M* value can be shown to be approximately 824 units 16.05% below the 982 target.

This may indicate as a lesson from this simulation that care should be taken with predictions far into the future with the current Bass model for CFD in greenhouses. What is however evident is the usefulness of the current Bass model in assessing CFD technology trends, also for greenhouse applications. As Bouwman and Van der Duin (2003) state it "scenario development starts where technology forecasting ends" (Bouwman and Van der Duin, 2003). This focuses the attention on the uncertainty aspect of forecasting illustrated in the previous

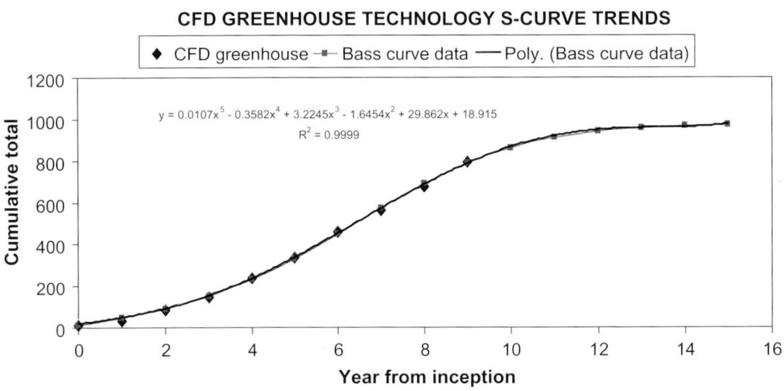

Figure 5: Technology trend for CFD for greenhouse applications.

Table 5: Optimized diffusion parameters for CFD greenhouse technology trend.

	M	P	Q
CFD technology	982.1213	0.025056	0.451653

illustration. The forecasts made, for example, for CFD technology in greenhouses may be usefully seen as starting points for trends and scenarios that may develop from that point onward (Table 5).

Trend for CFD Software Products

The CFD simulations of the flow environments for the greenhouse such as those presented in Fig. 4 may be achieved using different CFD software codes. These codes may be own in house developed codes or commercially available codes. The current flexibility and extensive development effort that have been put into commercial codes seem to favor the use of commercial codes.

The development trends for two CFD software products (the results of CFD technology) are now traced and compared via bibliometrics and simulated with the numerical Bass model to illustrate the extensive market penetration of as well as difference between two of these CFD codes. This is done for comparison purposes of the two different CFD codes as technologies in their own right.

The bibliometric data used to assess and compare the Bass curves shown in Figs. 6 and 7 are derived from research articles in Google Scholar (2008) with words, for example, "CFD Code A, year A to year B".

From Table 6 it may be inferred that research done with Code A may peak somewhat earlier than with Code B but with a steeper growth pattern. This may be due to the fact that Code A was introduced somewhat later than Code B. The maximum value should not be stressed too much in this case, as the initial data obtained are scarce and not as reliable.

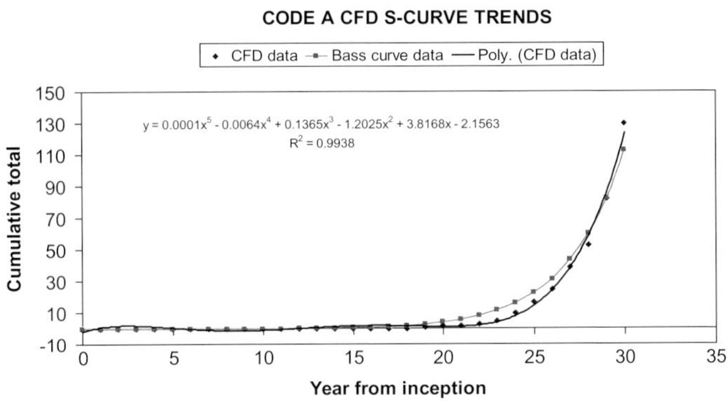

Figure 6: Technology trend for CFD Code A.

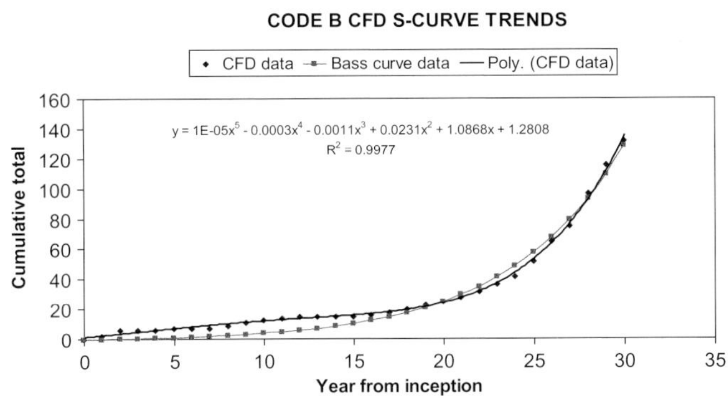

Figure 7: Technology trend for CFD Code B.

Table 6: Optimized diffusion parameters for code A and code B.

	M	P	Q
Code A	920.8454	2.26E-06	0.399259
Code B	2661.74	6.83E-05	0.176473

What should, however, be evident from Figs. 6 and 7 is the quite different curve forms in comparison for Codes A and B also a result of the diffusion parameters shown in Table 6. A competitive pattern of behavior may even be postulated from the difference in curve appearance for Codes A and B.

Discussion and Lessons Learned

From Figs. 2–5, it should be evident that it is possible to construct and model S-curves in an Excel spreadsheet environment using the Bass model that can effectively emulate or forecast observed cumulative publication data on CFD technology and CFD technology also in greenhouse applications. The observation is that the numerical and discretized version of the Bass model is appropriate over the time span involved.

A note of care should however be sounded when too much emphasis is placed on absolute values of forecasts. This is one of the lessons learnt in this research. There is uncertainty in data obtained as well as range of forecasts as have been shown for the greenhouse applications. This is where it is useful rather to think in terms of trends and scenarios as also emphasized by Bouwman and Van der Duin (2003). The insightful discussion on technology forecasting trends and scenarios for fuel cell technology by Daim *et al.* (2006) can also benefit this current research.

All the forecasting and subsequent analyses have assumed availability of information in the "market" (Bass, 2004) as well as no repeat buys. The effect of the introduction of one technology on another has thus not been addressed in the simulations.

Another lesson learnt is that it is very useful to be able to at least recognize the existence of a mature level, *M* of a technology at some point in the future as also emphasized by Tsai *et al.* (2014) in their discussion on the diffusion cloud computing technology where infrastructure and technology maturity played an important role. In the current research of CFD technology this is effectively illustrated in Fig. 5 where CFD in greenhouse simulation seems to approach some mature limit level that may indicate immanence of transfer to another or derivative technology as also referred to by Bae *et al.* (2007). By comparison the full field of CFD technology applications does not seem near saturation as indicated by Fig. 2.

All of these forecasts and trend casts for CFD technology have been made possible through the optimized regression of bibliometric data for CFD and the Bass diffusion model. This was done by the author making use of the nonlinear solver in a spread sheet application (Gottfried, 2007).

A trend cast is also added as a trend line $y = f(x)$ on the appropriate graphs in Figs. 2–7. As an example, the trend for the cumulative CFD technology "adopters" (publications) y as a function of year x shows a CFD technology trend:

$$\begin{aligned} y &= 0.002151x^5 - 0.091948x^4 + 1.476946x^3 \\ &\quad - 9.535003x^2 + 37.479869x - 1.585270 \\ R^2 &= 0.999819 \end{aligned} \tag{2}$$

Care should however be exercised in using these fifth order polynomials to forecast more than one or two time increments into the future as explained previously for the Bass equation. It can be shown that these fifth order polynomials can proceed to negative unrealistic cumulative future values under certain circumstances. This is also considered another valuable lesson learnt: Be careful when interpreting results from polynomial technology trend casts as opposed to from models such as the Bass model that contain more fundamentally reasoned assumptions in their development.

Conclusions

Some conclusions in this chapter were concerned with forecasting of CFD technology. It was shown that technology forecasting is possible for CFD although the issue is probably more trend casting than forecasting.

The Bass diffusion model was identified as a useful modeling vehicle for the CFD technology considered across the indicated time spans. One of the lessons learned in the process of simulation and forecasting is that care should be taken not to forecast more than one or two time units into the future at which time more data should be available and the current optimized Bass model can be updated for the CFD technology considered and further forecasting attempted. In essence, it seems better to do forecasting in steps of one or two time units into the future at which time the simulation should be updated with more data that has become available before new forecasts are attempted.

The application of CFD in the greenhouse was explored. It was demonstrated that circulatory flow patterns existed especially in the first two spans of the greenhouse considered. The use of CFD as technology to analyze the environment inside a greenhouse for design purposes has been effectively demonstrated. The diffusion of CFD technology in the context of analysis and to some extent design of greenhouse environments was specifically evaluated.

Future research may include the effect of one technology or derivative thereof on another technology applied to for instance the increased use of CFD for design applications in greenhouse design. The generalized Bass model or other models focusing on inclusion of competition (in the form of competing technologies) as well as the effects of geographical location may be studied for this application. Furthermore, alternative diffusion models such as Logistic/Pearl or the Gompertz models may also be studied for this CFD technology as indicated by Crompton (2001) in his study on the diffusion of new steelmaking technology. It may also provide additional insight to study the effect of multiple methods to model the growth of CFD technology. An approach similar to that of Daim *et al.* (2006) to combine the use of bibliometrics and system dynamics modeling for CFD technology forecasting may prove useful. This research may also be informed by the two stage system dynamics approach to ICT diffusion reported on by Chen *et al.* (2012).

Some initial research on a system dynamics approach to technology interaction and diffusion has been reported on previously (Pretorius *et al.*, 2013, 2015) but this can be extended to the area of CFD technology growth as well in future research.

References

Bae, J., Seetharaman, K., Suntharasaj, P., and Ding, Y. (2007). Technology forecasting of RFID by using bibliometric analysis and Bass diffusion model. *Proc. PICMET 2007*, 1637–1642, Portland, Oregon, USA, 5–9 August.

Bass, F. M. (2004). A new product growth model for consumer durables. *Management Science,* **50**(12): 1825–1832.

Bass curves for the diffusion of innovation (2008). http://www.useit.com/alertbox/basscurves.html, accessed 17/03/2008.

Bouwman H. and van der Duin, P. (2003). Technological forecasting and scenarios matter. *Foresight,* **5**: 4.

Chen, H., Wakeland, W., and Yu, J. (2012). A two-stage technology foresight model with system dynamics simulation and its application in the Chinese ICT industry. *Technological Forecasting and Social Change*, **79**(7): 1254–1267.

Clymer, N. and Asaba, S. (2008). A new approach for understanding dominant design: The case of the ink-jet printer. *Journal of Engineering and Technology Management*, **25**(3): 137–156.

Crompton, P. (2001). The diffusion of new steelmaking technology. *Resources Policy,* **27**: 87–95.

Daim, T. U., Rueda, G., Martin, H., and Gerdsri, P. (2006). Forecasting emerging technologies: Use of bibliometrics and patent analysis. *Technological Forecasting and Social Change*, **73**(8): 981–1012.

Danaher, P. J. Hardie, B. G. S., and Putsis Jr, W. P. (2001). Marketing mix variables and the diffusion of successive generations of technological innovation. *Journal of Marketing Research*, **38**(4): 501–514.

Dorf, R. C. and Byers, T. H. (2008). *Technology Ventures from Idea to Enterprise*. New York: McGraw-Hill.

Drejer, A. (1997). The discipline of management of technology based on considerations related to technology. *Technovation*, **17**(5): 253–265.

Google Scholar (Beta). (2008). http://scholar.google.co.za/, accessed 17/03/2008.

Gottfried, B. S. (2007). *Spreadsheet Tools for Engineers Using Excel.* New York: McGraw-Hill.

Incropera F. P. and De Witt, D. P. (2006). *Fundamentals of Heat and Mass Transfer*. New York: John Wiley.

Jordan, S. W. (2007). Technology forecasting with science indicators: The case of laptop battery futures. *"Proc. PICMET 2007"*: 1643–1650, Portland, Oregon, USA, 5–9 August.

Jørgensen, B. H. and Buckley, C. (2004). *Technology foresight and cleaner production in a highly polluted industrial area*. Denmark: Forskningscenter Risoe.

Kruger S. and Pretorius, L. (2007). The effect of ventilator configurations in naturally ventilated greenhouse applications. *Proceedings, UK National Heat Transfer Conference*, Edinburgh, 10–11 September.

Kruger, S. and Pretorius, L. (2008). Comparison of the indoor climate in multi-span and detached greenhouses with various ventilator configurations. *Proceedings ASME International Mechanical Engineering Congress and Exposition,* November 2–6, Boston, Massachusetts, USA.

Mahajan, V., Muller E., and Bass, F. M. (1995). Diffusion of new products: Empirical generalizations and managerial uses. *Management Science,* **14**(3) part 2 of 2: G78–G88.

Nieto, M. (2004). Basic propositions for the study of the technological innovation process in the firm. *European Journal of Innovation Management,* **7**(4): 314–324.

Ould Khaoua, S. A., Bournet, P. E., Migeon, C., Boulard T., and Chassériaux, G. (2006). Analysis of greenhouse ventilation efficiency based on computational fluid dynamics. *Biosystems Engineering,* **95**(1): 83–98.

Patankar, S. V. (1980). *Numerical Heat Transfer and Fluid Flow*. Washington, DC: Hemisphere.

Pouris, A. (2007). Nanoscale research in South Africa: A mapping exercise based on scientometrics. *Scientometrics,* **70**(3): 541–553.

Pretorius, L., Benade, S., and Kruger, S. (2008). Technology forecasting: The case of computational fluid dynamics (CFD). *Proceedings 4th IEEE International Conference on Management of Innovation and Technology*, 21–24 September, Bangkok, Thailand.

Pretorius, L., Benade, S., and Kruger, S. (2011). Technology diffusion and forecasting: The case of computational fluid dynamics (CFD) for simulation of greenhouse internal environments. *International Journal of Innovation and Technology Management*, **8**(1): 27–39.

Pretorius, L., Pretorius, J. H. C., and Benade, S. J. (2013, July). A system dynamics approach to technology interaction including cyclic behaviour. In *Technology Management in the IT-Driven Services (PICMET), IEEE 2013 Proceedings of PICMET'13*, pp. 2171–2180.

Pretorius, L., Pretorius, J. H. C., and Benade, S. J. (2015). A system dynamics approach to technology interaction: From asymptotic to cyclic behaviour. *Technological Forecasting and Social Change*, **97**: 223–240.

Shang, J. S., Li, S., and Tadikamalla, P. (2004). Operational design of a supply chain design system using the Taguchi method, response surface methodology, simulation and optimization. *International Journal Production Research*, **42**(18): 3823–3849.

StarCCM+. (2006). *CFD Manuals*, CD-Adapco.

Sturgess, G. J. (1996, November). Combustor design trends for aircraft gas turbine engines. In *ASME 1996 Turbo Asia Conference*, pp. V001T02A007–V001T02A007. American Society of Mechanical Engineers.

Tsai, J. M. and Hung, S. W. (2014). A novel model of technology diffusion: System dynamics perspective for cloud computing. *Journal of Engineering and Technology Management*, **33**: 47–62.

Van der Merwe, L. (2002). A product development process for a photovoltaic water pump system in a small to medium enterprise. Doctoral Thesis. Rand Afrikaans University, South Africa.

Van Staden, M. P., Pretorius L., and Meyer, J. P. (1998). Simulation of heat exchange in large air cooled condensers. *Proceedings of the 11th International Heat Transfer Conference*, **6**: 155–160, Kyongju, Korea, 23–28 August.

Versteeg, H. K., and Malalasekera, W. (2007). *An introduction to Computational Fluid Dynamics: the Finite-Volume Method*, 2nd ed. Harlow: Pearson, ISBN 978-0-13-127498-3.